PLANNING BY LAW AND PROPERTY RIGHTS RECONSIDERED

Planning By Law and Property Rights Reconsidered

Edited by

THOMAS HARTMANN
Utrecht University, The Netherlands

and

BARRIE NEEDHAM
Radboud University Nijmegen, The Netherlands

Routledge
Taylor & Francis Group
LONDON AND NEW YORK

First published 2012 by Ashgate Publishing

Published 2016 by Routledge
2 Park Square, Milton Park, Abingdon, Oxfordshire OX14 4RN
711 Third Avenue, New York, NY 10017, USA

First issued in paperback 2016

Routledge is an imprint of the Taylor & Francis Group, an informa business

British Library Cataloguing in Publication Data
Hartmann, Thomas.
Planning by law and property rights reconsidered.
1. Right of property. 2. Land use—Law and legislation.
3. City planning and redevelopment law.
I. Title II. Needham, Barrie.
346'.045–dc23

Library of Congress Cataloging-in-Publication Data
Hartmann, Thomas.
Planning by law and property rights reconsidered / by Thomas Hartmann and Barrie Needham.
p. cm.
Includes index.
ISBN 978-1-4094-3721-5 (hardback : alk. paper)
1. Right of property. 2. Land use—Law and legislation. 3. City planning and redevelopment law. I. Needham, Barrie. II. Title.
K721.5.H37 2011
346.04'5–dc23

2011048029

ISBN 13: 978-1-138-25694-1 (pbk)
ISBN 13: 978-1-4094-3721-5 (hbk)

Contents

List of Figures and Tables

Figures

Tables

List of Contributors

Stephen Berrisford is an independent researcher and consultant based in Johannesburg working on planning law and policy reform in Sub-Sahara Africa. He qualified as a lawyer and an urban planner from the University of Cape Town and holds an MPhil by research from the University of Cambridge (thesis title 'Property rights and Urban Planning in South Africa'). Berrisford worked for four years at a high level in the South African government responsible for the implementation of the first post-apartheid planning legislation. His consultancy work is split between South Africa and other countries in the region with particular experience of the planning and urban law reform processes in Zambia, Ethiopia, Kenya and Lesotho. He works closely with Urban LandMark ('making urban land markets work better for the poor in Southern Africa') and the African Centre for Cities at the University of Cape Town.

Edwin Buitelaar works as a senior researcher and programme leader at the PBL Netherlands Environmental Assessment Agency, a government-funded research agency, and as a research fellow at the Amsterdam School of Real Estate of the University of Amsterdam. He does research on land development, land-use planning, planning law and (social) housing. Before he moved there at the end of 2007, he worked as a lecturer (assistant professor) and a researcher in the department of Spatial Planning and the research team Governance and Places (GaP) of the Radboud University Nijmegen.

Esther Geuting is a partner of a consultancy firm (Stec Groep) in land and property markets in the Netherlands. She does research and advises on land markets, the value of land and real estate, restructuring of neighbourhoods and industrial estates, and property development strategies. She has been a part-time researcher in the department of Spatial Planning and the research team Governance and Places (GaP) of the Radboud University Nijmegen. In 2011 she finished her PhD thesis on the subject of market structuring as a policy tool in planning.

Thomas Hartmann is currently a post-doctoral researcher at the Department of Human Geography and Spatial Planning, Utrecht University. In 2009, he finished his PhD on *'Clumsy Floodplains – Responsive Land Policy for Extreme Floods'* at the department for Land Policy, Land Management and Municipal Geoinformation, TU Dortmund. Before, he studied Spatial Planning at the School of Spatial Planning, Dortmund. In his research on environmental governance he combines engineering, socio-political research and anthropological studies to explore the relationship between land and water. Since 2008, Dr. Thomas

Hartmann is an office holder in the International Academic Association on Planning, Law, and Property Rights.

Harvey M. Jacobs, PhD, is on the faculty of the University of Wisconsin-Madison, USA where he holds a joint appointment as Professor in the Department of Urban and Regional Planning and the Gaylord Nelson Institute for Environmental Studies. Professor Jacobs has been a member of the faculty since 1984, and has served as Chair of the Department and Director of the University's Land Tenure Center. Prof. Jacobs is the editor of *Private Property in the 21st Century: The Future of an American Ideal* (Edward Elgar, 2004) and *Who Owns America? Social Conflict Over Property Rights* (University of Wisconsin Press, 1998), and the author or co-author of approximately 100 professional articles and essays, of which over 30 focus specifically on the social and legal aspects of property rights. Throughout his career Professor Jacobs has won awards and recognition for his research and teaching.

Leonie B. Janssen-Jansen is an associate professor at the Faculty of Social and Behavioural Sciences, University of Amsterdam, research institute AISSR (Amsterdam Institute for Social Science Research). She was elected as President of the International Academic Association on Planning, Law and Property Rights in February 2010. She is also one of the founders of this worldwide association. She holds a PhD in Urban and Regional planning from the University of Utrecht (2004) and a Master degree (cum laude) in Urban and Regional planning (1998), also from the University of Utrecht. Her post-doctoral research, financed by the Dutch Scientific Organization, focused on non-financial compensation instruments in planning. Her current teaching, research and publications focus on urban development, real estate, urban transformation, regional planning, growth management, regional governance, non-financial compensation instruments, regionalization, institutional planning and planning and law.

Erwin van der Krabben holds a position as professor of land management and area development in the Institute of Management Research, Department of Spatial Planning at Radboud University Nijmegen. He also holds a position as (part-time) professor of real estate in the School of the Built Environment, University of Ulster (Northern Ireland). He has published in a variety of academic journals with respect to land policy, area development, and the regulation of land and property markets. He has acquired substantial funds for scientific research (including Transumo, Habiforum, NICIS and COST) and supervises a number of PhD projects in those academic fields. He is the coordinator of the research program Governance and Places (group of circa 50 researchers). In an international context, he is AESOP country representative for the Netherlands and co-coordinates the EU COST Action Land Management and Urban Dynamics. Erwin van der Krabben is frequently invited to advise or give expert opinions to various organizations (including the Ministry of spatial planning, provinces, municipalities).

Barrie Needham (1942) studied economics at the University of Cambridge and worked in practice and academia in England before taking an appointment at the Radboud University of Nijmegen in 1978. He was employed in the Department of Spatial Planning there, becoming professor in 1994, until his retirement in 2007. His main research interests and publications concern the relationship between land-use planning and land policy. His latest books are *Planning, Law And Economics* (Routledge, 2006), *Dutch Land-Use Planning* (Sdu publishers, The Hague, 2007), *Planning van bedrijventerreinen* (The Planning of Industrial Estates – together with Louw, Olden and Pen – Sdu publishers, The Hague, 2009).

Mark Oranje is Professor and Head of the Department of Town and Regional Planning at the University of Pretoria. He has authored and co-authored numerous academic papers, articles, chapters in books and technical reports. His key areas of interest are planning policy, long-term planning and planning history. Over the last two decades, he has acted as a consultant to a number of national and provincial government departments, national and provincial planning commissions and science councils on issues relating to municipal integrated development plans, local and regional economic development, municipal, provincial, national and supra-national planning, land-use management, intergovernmental relations and the integration of land use and transport planning.

Tejo Spit (1955) got his Master's degree as a physical planner at the Nijmegen University (Department of Urban and Regional Planning) in 1983. Since then he has been working both in the academic world (Radboud University of Nijmegen and University of Utrecht) and the more problem-oriented world of municipalities (first the municipality of Enschede and from 1985 onwards the Association of Netherlands Municipalities). In that way, he has developed wide knowledge of spatial planning in (local) practice. In this period he also wrote his thesis on the subject of fiscal austerity and urban innovative policies (1993). In 1995 he moved to a full time academic career as an associate professor at the Department of Human Geography and Planning (Faculty of Geosciences, Utrecht University). He is now full professor in Urban and Regional Planning and chairing the department. He is specialized in land policy, planning methodology, infrastructure planning and administrative aspects of spatial planning. He lectures on these subjects regularly within the Netherlands and abroad.

A. Dan Tarlock is Distinguished Professor of Law at the Chicago-Kent College of Law, Chicago, USA. He holds an AB and LLB from Stanford University and has previously been a permanent member of the faculties of the University of Kentucky and Indiana University, Bloomington. He has also visited at several law schools all over the USA. His teaching and research interests include water law, land-use controls and international environmental law. He has lectured in 15 countries. He is a co-author of a leading United States land-use controls casebook, and he is a member of the special legal advisors to the Submissions Unit of the

Commission on Environmental Cooperation in Montreal, Canada. In 2009, he was appointed to a standing committee on the National Research Council to provide strategic planning advice to the US Army Corps of Engineers.

Joost Tennekes is researcher at the PBL, Netherlands Environmental Assessment Agency, the advisory body of Dutch government on issues of spatial planning, nature and environment. He published on the impact of increased flood risk on urban planning and design, the use of risk zoning instruments in the Netherlands, and on the adaptability of the Dutch spatial planning system to climate change in general. His background is in Public Administration (PhD University of Twente) and Philosophy of the Social Sciences (MA Free University in Amsterdam).

Huub Ploegmakers is a PhD student at the Department of Spatial Planning, Radboud University Nijmegen. He joined the Department in October 2008. He obtained his bachelor and master degree in Spatial Planning at the same University. As a PhD student he is involved in a NICIS funded research programme concerning (re)development of industrial estates. His research focuses on redevelopment of industrial estates in relation to the functioning of the industrial land market.

Michael Wegener is presently a partner in Spiekermann & Wegener Urban and Regional Research, Dortmund, Germany. Until 2003, he was Professor and Director of the Institute of Spatial Planning at the Faculty of Spatial Planning of the University of Dortmund, and from 1988-89 Professor at the Department of Civil Engineering of the University of Tokyo, Japan. He graduated in Architecture and Urban and Regional Planning at the Technical University of Berlin and received his PhD degree (Dr.-Ing.) at the RWTH University of Technology, Aachen. His main research fields are planning theory and theories and models of spatial development, in particular urban and regional development, European urban systems and transport networks, the land-use transport interface in cities and regions, and the regional impacts of European transport infrastructure.

Acknowledgements

We would like to thank Professor Dr Willem Salet for his kind and supportive feedback and review. Willem Salet is Professor of Urban and Regional Planning at the Faculty of Social and Behavioural Sciences, University of Amsterdam.

In addition, we would like to thank Professor Rachelle Alterman for supporting the idea of publishing this book. Rachelle Alterman is Professor at the Faculty of Architecture and Town Planning at the Technion – Israel Institute of Technology in Haifa, and founder of the International Academic Association on Planning, Law, and Property Rights (see: www.plpr-association.org).

The International Academic Association on Planning, Law, and Property Rights provided a good forum for discussing the ideas in, and drafts of, the contributions of this book.

Introduction
Why Reconsider Planning By Law and Property Rights?

Thomas Hartmann and Barrie Needham

Present day spatial planning began with something called urban planning, or land-use planning, or town-and-country planning. These were introduced in the second half of the nineteenth century in response to the appalling conditions in the towns and cities which were growing rapidly to accommodate the industrial revolution. Governments decided that the use of property rights in land and buildings should be regulated. Owners should neither always be allowed to build what they wanted where they wanted, nor always to use their buildings in ways that they wanted. So regulations were introduced, and plans were made to guide and legitimize the contents of those regulations. Later, in the first half of the twentieth century, such regulations were extended in response to cities growing rapidly into the countryside. So, gradually, most industrialized countries came to have a system of plans, control over building development, and ordinances about environmental nuisances. Gradually, spatial planning has grown as a recognizable and recognized discipline – academically and in practice. Now, the task of planning is much wider than the facilitation and regulation of urban growth. The tasks and challenges of planning have changed over the last decades – and they keep changing. As planning challenges change, planning approaches also change.

Nevertheless, those approaches are still rooted in the activities of making, implementing and enforcing legal rules about how people may use their property rights over land and buildings. This we call planning by law and property rights and that is the subject of this book. It is still commonly practised, for it includes the making of land-use plans, the issuing of building permits, and so on. But is it necessary to write a book about it?

The answer is yes, for it is our contention that planning by law and property rights is so familiar and taken for granted that we do not think about the theory behind it. As a result, we do not think abstractly about its strengths and weaknesses, about what can be achieved with it and what not, how it can be improved, how it could be complemented. Such reflections are essential to cope with current and future challenges of spatial planning. In this book, we make the (often implicit) theory behind planning by law and property rights explicit and relate it to the challenges. Therefore, we ask questions such as: Is planning by law and property rights desirable? What are its strengths and limitations? Because of its limitations, could it be replaced by another way of planning? Or is it indispensable, in one form of another? In short, we 'reconsider' planning by law and property rights.

Our argument is that planning by law and property rights is unavoidable and indispensable in a society with a rule of law. But it is not just a necessary evil, for it can give the users and producers of the physical environment the stability and certainty that they desire. On the other hand, it has its limitations, for the corollary of stability and certainty is inflexibility. Once rights in land have been determined by a combination of private law and public law, changing them can be difficult. If the assignment of property rights thus determined turns out to be inconvenient or out-dated, spatial planning by legal regulations carries the danger of creating lock-in situations. Planning by law and property rights which aims to be flexible runs the risk of doing that by treating peoples' property rights in an ad hoc way.

It is partly because of those limitations that in the last few years planning theorists have given much attention to another way of spatial planning, namely planning by agreement or 'collaborative planning'. This book is not about such planning, but we do have to pay some attention to it as a possible alternative, or complement, to planning by law and property rights, as a way of remedying the limitations of that latter.

This introduction analyses a central dilemma with this type of planning, namely how can planning by law and property rights be practised to the best advantage without creating undesirable (for example unsustainable or socially unfair) lock-in situations? In the remainder of this book, this and other important issues for the development of such a theory of planning by law and property rights are explored.

Planning, Law, and Property Rights

> Every natural or legal person is entitled to the peaceful enjoyment of his possessions. No one shall be deprived of his possessions except in the public interest and subject to the conditions provided for by law and by the general principles of international law. The preceding provisions shall not, however, in any way impair the right of a State to enforce such laws as it deems necessary to control the use of property in accordance with the general interest or to secure the payment of taxes or other contributions or penalties. (Article 1 of the Protocol of the European Convention on Human Rights)

In order to give persons that 'peaceful enjoyment', the state creates property rights. The owners of land (or of rights in land) can ask the courts to protect them if others endanger their 'peaceful enjoyment'. Property rights are created under what is called 'private' or 'civil' law: the law is enforced only if a private person (legal person) takes the initiative. That person considers that his/her interest is being damaged by the action of another legal person (which could be a government body), and asks the state to correct the situation. Those laws are called 'private', because they regulate the direct relationships between legal persons: the state acts only if requested to do so. Examples of private laws concerning land are the laws about freehold and leasehold interests, about easements and covenants, about mortgages, and so on. Rights give power to the owner of the rights, power not only over the object (or circumstances) of interest, but also over all others, who may be excluded from that right (Bromley 2000: 29–30). 'A right', he explains, 'is to have the ability and the capacity to require the government to act in your behalf' (Bromley 2000: 29).

In these terms, it would seem that planning 'interferes' with the peaceful enjoyment of property. For under land-use planning, a state body says what the owner of a property right in land or buildings may or may not do with his or her land. The state body might even be empowered to acquire that right compulsorily (expropriation). The state is given those powers under what is called 'public' or 'administrative' law. Such laws are not only made, but also enforced, by the state: it is the state which intervenes if a law is broken. Such laws are called 'public' because they concern directly the relationship between the citizen and the state. Examples of public laws concerning land are the laws for land-use planning, much of environmental law, many laws concerning roads and highways, and so on.

In both cases – private and public law – we are talking about formal laws, laws which are enforceable in courts. They are all determined by the state, for it must be possible to enforce them, and the state has the monopoly on the legitimate use of power (for example levying fines, imprisoning, requiring compensation for damage).

Three things follow from this. The first is that all formal laws are a social creation (Bromley 2000: 28). Bromley explains that the law 'reflects the influence and interests of yesterday and, through its application, the law seeks to bind the future' (Bromley 2000: 29). Seen in this way, whenever a spatial plan or a planning measure specifies how a particular plot of land may or may not be used, it is socially constructing and assigning property rights by using the law (Needham 2006: 4). The second is that in a state with a rule of law, all public actions to 'interfere' with the exercise of property rights must be justifiable by law (no arbitrary actions). This means that all land-use planning is inevitably 'planning by law and property rights'. The third is that there must be a close connection between the use of private and of public law, for both affect how land is used, and the two types of law may not be used in a contradictory way on one plot of land. For example, if public law intervenes in the exercise of an existing property right, that 'contradiction' must be carefully and explicitly resolved by legal means.

No matter how many levels of planning are involved in preparing planning decisions, no matter how many stakeholders have contributed to a decision, no matter which parties and lobby-organizations have influenced a decision, in the end, a private landowner is either restricted or enabled to enjoy his possessions – to use the term of the European Convention on Human Rights. It makes 'a person poorer or richer, by affecting the value of the rights on her land' (Needham 2006: 3). For that reason, planning by law and property rights requires a thorough understanding of the law (Ingram and Hong 2009: 3): spatial planners need to reflect continually on the legal implications of their plans. The sought-for understanding, however, is elusive, since neither planning nor law and property rights are the same in all places and at all times. The relation between political power – which spatial planning to some extent exercises – and property is described by Goodin and Tilly as context dependent: 'as analysts of political processes, we have no choice; we must place rights to resources in context' (2008: 6). The reflection on the context provides an opportunity to rethink how spatial planning can work constructively with law and property rights, precisely so as to achieve the chosen planning aims better – which could mean fairer, more efficient, more effective or sustainable plans. There is an unavoidable logical relationship between planning, law, and property rights, and certain aspects of this need to be worked out better.

Public and Private Interests

'Rights in land are granted by the political unit because of the larger social benefits to arise therefrom' (Bromley 2000: 28). 'Defenders of private ownership of land argue that it promotes individual liberty, political stability, and economic prosperity' (Ellickson 1993: 1317). What is then the purpose of imposing *in addition* planning regulations on how private people use their property?

The ultimate purpose is to achieve a way of using land which is 'better' than if the private person had been allowed to determine that land use in the absence of the plan, even though working within the constraints of private property rights. But better for whom?

It can be – is often – better for others than the property owner. The necessity of governmental intervention in private property results from the inevitability of scarcity which creates conflicts (Bromley 2000). Land is a scarce good; a decision for one use often excludes other uses. For example if landowners along a river are enforced to leave their land empty so as to retain the water of the next flood wave, landowners downstream profit from that, whereas the upstream landowner cannot use his/her land (or only within crucial constraints). Sometimes these 'others' who gain from planning are few and easily identifiable (for example if actions can be taken to prevent someone causing nuisance – an external negative effect – to the immediate neighbourhood). Sometimes these others are very many and difficult to identify and isolate (for example diffuse negative external effects, such as when a beautiful landscape would be damaged by the private action). 'Better' can also even refer to the private person owning the land which is regulated: for example, planning can make it possible to redevelop a difficult site profitably (a positive external effect).

Spatial planners work within that framework: 'the planner is the leader, the professional' (Schiller 2001: 10). 'Traditional land-use planning … aims at controlling land use' (Albrechts 2004: 744). However, it is not only the owners of rights in land who are affected: there can be other stakeholders too, who have an interest in the use of space. For example, water managers, environmental agencies, or other public authorities. All those stakeholders might perceive the regime of law and property rights as a threat or a constraint to achieving their cherished design (Schiller 2001: 11).

Planning By Law and Property Rights

Two ways of planning by law and property rights can be distinguished. The first way is the most familiar: it is when a public agency makes plans using public law about how land may be used. That plan sets the conditions under which people may use their property rights. That use of public law should not be an 'uninspired instance of bureaucratic business', as Forester (Forester 2004: 242) criticizes much planning practice, for it can also be applied creatively in such a way that it directs and steers certain uses of private law. It could include planners not trying to control, but to create a frame for stakeholders' activities. This could mean determining the desired outcome and leaving it to the stakeholders how they achieve that; that is, not to try to determine the *procedures* but the *substance* of an intended future land use. The second way of planning by law and property rights is more indirect. It consists of a public agency creating markets in property rights, purposefully so that the outcome of those markets is a physical environment which the public agency wants to achieve. Creating transferable development rights is a well-known example of this, or making it possible for the residents of a neighbourhood to set up a residents' association which maintains the qualities of that neighbourhood. This kind of spatial planning works by structuring markets in property rights.

With both approaches, plans need to remain within the legal boundaries specified by both law and property rights. But planners can nevertheless think constructively about what are appropriate property arrangements for certain locations. For example, who should own public spaces, who the gardens in front of the houses, who the streets, who the retention area upstream, who the nice view of a landscape? More generally, planning can be regarded as spatial governance: how should the use of space be 'managed'?

Planning: Incomplete, Experimental, Complex, Normative

Spatial planning is always incomplete, experimental, complex and normative (for example Baum 1977; Bertolini 2010; Healey 2009; Hillier 2010). It can be seen as a spatial policy experiment: 'policy changes are experiments based on more or less informed expectations about potential outcomes and the distribution of these outcomes for participants across time and space' (Ostrom 2007). Planning is in other words always experimental (Hillier 2010: 14). In addition to that, this 'complexity of the ever-changing biophysical world combined with the complexity of rule systems means that any proposed rule change faces a non-trivial probability of error' (Ostrom 2007). Ostrom concludes that this places a heavy responsibility on authorities who decide on the allocation and distribution of goods. On top of that, these authorities cannot rely on simple models and facts to predict the effects of their planning decisions, for planning is 'characterized by (among other factors) fragmentation, plurality, subjectivity and interpretation' (Allmendinger 2002: 88). 'Planners are always thinking about and making plans for possible futures while continually being dragged back to the messiness of 'going on in the world'. Planning is thus always incomplete' (Hillier 2010: 13). But it is expected of planners that they 'balance and combine the alternatives in order to achieve the best result' (Davy 1997: 247). They are expected to achieve fairness of distributional effects besides economic efficiency (Needham 2006: 139; Davy 2005: 117–119). Bertolini put the challenge for planning like this: 'The central challenge of planning in the face of irreducible uncertainty is to acknowledge deep uncertainty and simultaneously provide operational policy recommendations' (Bertolini 2010: 93). So we can conclude that planning is incomplete, experimental, complex and normative.

> The streets have been paved, and roads now connect all places; houses shelter virtually everyone; the dread diseases are virtually gone; clean water is piped into nearly every building; sanitary sewers carry wastes from them; schools and hospitals serve virtually every district; and so on. The accomplishments of the past century in these respects have been truly phenomenal, however short of some persons' aspirations they might have been. But now that these relatively easy problems have been dealt with, we have been turning our attention to others that are much more stubborn. (Rittel and Webber 1973: 156)

In accordance with the famous paper on 'Dilemmas in a general theory of planning' (Rittel and Webber 1973), we name this 'wicked planning'.

Planning law and property rights provide an orientation in this wicked planning: 'The legally binding nature of most EU land-use plans provides legal certainty' (Albrechts 2004: 745). This is a strength of law and property rights. The legal system provides a reliable and predictable framework for landowners. The zoning plan (US), development plan (UK), the Bebauungsplan (GER), the bestemmingsplan (NL), or whatever it is called in different countries, provides a legal basis for landowners to develop their land – to peacefully enjoy their possessions. This is so whether one considers that these plans create a development right which the owner did not have before, or whether one considers that the owner always has a development right which the plan can prevent him from exercising. The plan can also regulate the use of rights by other – possibly competing, possible complementary – landowners, and in that way too provide a predictable framework for the landowner.

Changing Rights is Difficult

The crucial point with rights in land use is that, once they are assigned, the hurdles to removing or changing them are high because of the constitutional protection of property rights. The rights which are assigned by planning are usually granted freely, but if the same institution that assigns the rights wants to have them back, it costs money. If a governmental intervention is regarded as an expropriation, exercising eminent domain, or compulsory purchase (depending on the respective legal system) landowners are entitled to compensation. If the restriction on the use of a right is regarded as a taking (see for example Cooter and Ulen 2004: 175), then likewise compensation might be required. So it can be difficult to change plans, because that can entail paying compensation or give rise to legal objections. This creates a tension, which Albrechts describes in the continuation of the quote above like this: 'land-use plans provide[s] legal certainty but make[s] the plans far more rigid and inflexible and less responsive to changing circumstances' (Albrechts 2004: 745).

In the face of wicked planning, this inflexibility is a problem. If an assignment of property rights or a certain rule for using land leads to an inconvenient allocation and distribution of goods, change is difficult. Subsidizing private building activity and providing plots for single family houses on greenfield sites was an important planning task in the past: planners now seek regulations to cope with shrinking cities or land thrift (Davy 2009, 2010). In the 80s, planners tried to realize the car-friendly city: now planners and policymakers try to keep cars out of the city centres, for example with tolls (like in London). Spatial planners have permitted building on riverside property: now, they try to restrict such developments since they have become inappropriate in the face of increasing flood risk. It cannot be predicted whether or not the assignments of certain rules to the land will be in line with what future planners want to achieve. Brian Arthur described from an economist's perspective how technical developments can become locked-in through 'random events' in the course of time (Arthur 1989: 1). Social situations also can become locked-in (Thompson 1997: 214). Such lock-in situations are situations, in which a better (more efficient, more sustainable, more desirable …) situation will not be achieved due to an entrenchment in the current system. The current system maintains itself, and the situation will not change without an 'optimal perturbation' (Thompson 1997: 213–214).

Particularly in countries where, for example due to colonialism or revolution, a new legal system has been set up, lock-in situations can become crucial. After the German reunification, the (West) German Grundgesetz (Constitution) was adopted in Eastern Germany and this accepted the socialistic land reform of 1945–1949. This land-political act of the reunification of Germany is a very sensitive and highly disputed issue, and it is also an example of an entrenchment of a spatial order (Paffrath 2004). This can happen in other countries too. Oranje and Berrisford discuss the tension between the recently introduced planning law in South Africa and the entrenchments of a certain social order (see Berrisford and

Oranje, Chapter 3). In addressing these issues, the highly political character of legal planning decisions is highlighted.

For planning by law and property rights, this means that the inflexibility of assigned rights on the one hand, and the wickedness of planning on the other hand, might lead to inconvenient rules and property arrangements. An optimal perturbation would cost a lot of money (and effort and political willingness). Michael Thompson recommends trying to avoid such lock-in situations by switching the focus 'from anticipation to inflexibility reduction' (Thompson 1997: 214). What does that mean for spatial planning and the role of planning regulations? Should the assignment of development rights in this regard then only be temporary? How can law and property rights play their stabilizing and certainty-producing role in the light of this? These are questions, which are not easy to answer, but which cannot be ignored.

Towards Less Wicked and Less Inflexible Planning

'A governance structure that is based on private property rights enforced by external authorities is not always necessary or optimal. Users are often capable of nurturing trust and reciprocity to solve their collective action problems' (Ingram and Hong 2009: 6). Accordingly, in recent years and decades, planning by public regulation of land use seems to be losing support (Wegener 1999). Some even dismiss it as being command and control (whereas planning can forbid but not command!), or as top-down and centralized (Selle et al. 2000).

Planning theorists and practitioners have developed in the past years and decades alternative ideas, approaches, and concepts to remedy the purported limitations and defects of planning by land-use regulation. The theoretical attempts towards a new spatial governance are often referred to with 'post-isms', as for example 'post-positivism' (Allmendinger 2002), or 'post-industrialism' (Baum 1977). Most such planning theories claim that 'conventional planning' – such as conducted in the decades after the Second World War – needs to change. Conventional planning refers here to the approach which regards planning as an engineering task, achieved through the application of law and regulations. It regards law and property rights in a very procedural perspective (in contrast to the substantive aspects of planning by law and property rights). Even those who might not agree that conventional planning is out-dated and should be rejected for the above reasons, recognize that planning by regulating property rights is not effective in all situations; for example because of scarce public resources, or the limits to controlling the spatial activities of stakeholders (Selle 1994).

Referring to the lock-in trap mentioned above, two types of planning approaches are of interest: those which reduce inflexibility of planning, and those which try to make planning problems less wicked.

Attempts which address the inflexibility of conventional planning often propose competitive and experimental approaches to planning. The recognition that planners are just 'fallible advisors who operate like everybody else' (Allmendinger 2004: 88) led to more flexible and adaptable planning schemes; Michael Wegener describes the change as 'from grand designs to small, manageable, and correctable incremental improvements' (see Wegener, Chapter 8).

It is often argued that more involvement of 'the market' would make planning more flexible. One form of this argument is for less planning by public agencies, leaving more room for market actors to determine the form and use of the physical environment. This would be more flexible because public agencies then determine much fewer details about the physical environment and because market actors can react more quickly to external changes and to unexpected consequences. Planning by law and property rights which works by structuring markets in property rights (see above) is one version of this 'more market'. Neo-liberalism goes much further, for it requires that public authorities withdraw from formulating what they would like the physical environment to become. When Thatcherism in Britain propagated neo-liberal planning, this was inherently hostile towards the idea of (conventional) planning (Healey 2003: 102). Another version of 'more market' is the trend in European Urban Policy, where strengthening regional competitiveness has become a key issue (Frank 2006: 51). Market schemes – this is the underlying belief – are more flexible and adaptable to new developments.

Another way of involving 'the market' is by close cooperation between public and private partners. Here, public authorities do not withdraw from spatial planning, but get even more closely involved by changing the physical environment directly. However, they do this in collaboration with market actors. One of the arguments for doing this is that it reduces inflexibility by giving the market actors a broad frame or rules within which to work flexibly and to adapt where necessary. The success story of public-private partnerships in municipal planning in the 1990s (Heinz 1993) or project-oriented planning experiments like during the IBA Emscher-Park (International Building Exhibition) in the 1990s (see for example Kurth et al. 1999) mark attempts to cope with the inflexibility of planning by law and property rights

It seems that planners have lost confidence in top-down hierarchical planning, and either want to withdraw and leave more tasks to the market, or they want to pursue voluntary cooperation, searching for consensus in a multi-level governance. Planning experiments and trial-and-error approaches have become a common method for gradually adapting plans to new challenges. For planners, the ability to adapt – or correct – plans has become more important in the last years. Bertolini proposes the concept of evolutionary planning as a response to uncertainty and complexity in spatial planning (Bertolini 2010: 83). This concept is based on the idea of gradual changes to current systems, which adapt in that manner to changed environments.

The other planning approach is to make planning problems less wicked. This could be done, it is often argued, by more participative and collaborative schemes to increase the deliberative quality of planning. For example 'communicative planning' promotes planning which minimizes inequalities of power and knowledge by genuine participation (Huxley 2000: 369). Selle advises that planning should be carried out less by law and property rights and more by consensus and cooperation (see for example Selle et al. 2000). Social contexts of planning became important during and after 'post-ism' paradigm shifts in spatial planning: 'land rules arise not so much from law as from customary norms that are enforced through diffuse social sanctions' (Ellickson 1993: 1319).

We observe a trend towards planning approaches outside the realm of planning law and property rights in order to reduce uncertainty and complexity: 'The traditional model of government setting the framework for private decisions has been replaced by a flexible system of communication and adjustment between public and private actors called governance' (Wegener, Chapter 8). To what extent are such 'collaborative' schemes able to cope with inflexibility and wicked planning problems?

Today: Collaborative Planning

The label 'collaborative planning' has been used for various inclusive participative governance processes in project design and development. But 'the metaphor "collaborative planning" became used and misused' (Healey 2003: 108), Healey complains. Groups, politicians and policymakers used the term in order to promote certain participative policy-designs. But collaborative planning is more than just participation. It can include also collaborative implementation. Even wider, it provides a certain lens (Forester 2004) to look at planning. It regards planning as an interactive governance process, which can occur in many modes; this governance process is 'shaped by wider economic, social, and environmental forces that structure, but not determine, specific interactions' (Healey 2003: 104). Regarded in this way, collaborative planning stands for planning which is wider than planning by law and property rights. But is it the answer to the established weaknesses in that latter? In our reflection on planning by law and property rights, we need to recognize that collaborative planning schemes also are constrained and can carry threats.

A very close collaboration between public and private stakeholders, but also between different public stakeholders, is not always in the interests of openness and democracy. Collaborative planning is often treated as an achievable mode of governance. 'Ironically the progressive credentials of spatial planning in terms of consensus building, policy integration, and the search for "win–win–win" solutions may have helped script out oppositional voices' (Allmendinger and Haughton 2010: 803): and that removes the legitimacy of the planning (op. cit. 2010: 804). In particular, the incorporation of private entities in collaborative planning schemes contains the threat to 'sell out the public interests'. If public agencies rely too much on models like public-private partnerships or getting private actors to take over public tasks, commercial issues might drown public interests, risks and burdens might be unequally distributed between the public and the private parties (Heinz 1993), or public spaces might be changed into club goods, available only to members of the 'club'. So, recent trends in planning – aiming at more integrative concepts and collaboration between various stakeholders – make planning a political process (Stein 1995: 393) and thus move it from being part of the executive 'estate' (administrators implement rules made by the legislator) to being part of the legislative 'estate' (administrators themselves make the rules), or to a mixture of both (Durner and Ludwig 2008). This raises the question of legitimacy: how can the discourse on planning by law and property rights deal with this?

Another constraint of collaborative planning comes from the Habermasian assumption that communication inherently contains the possibility of consensus (Huxley 2000: 370). This leads to the 'mental shortcut' (Billé 2008: 77) that collaborative schemes are the ultimate planning tool, which can solve complex solutions. However, conflicts are not automatically solved just by bringing all stakeholders to a roundtable and expecting them to discuss the problem until it is resolved (Billé 2008: 77; Huxley 2000). Informal public discourses 'can come to discover or create a common good' (Fraser 1990: 72), but do not formally make decisions about that public good (Huxley 2000: 370). In addition, even if a consensus about a planning issue has been found by collaborative planning, the open discourse which should lead to that achievement 'emerging' is already a weakness, since the consensus is not necessarily a desirable end-state in itself (Huxley 2000: 372). This means that a consensus which emerged from an open discourse can easily be disturbed by the discourse itself – if the discussion opens up again or a new stakeholder with a new opinion joins the discussion. Collaborative planning which focuses on non-regulative instruments for plan implementation (for example round-table discussions, egalitarian schemes, or market approaches) might not be able to withstand real land-use conflicts, and in that way it can create rather instable (and thus unsustainable) solutions. So, collaborative planning based on communication and deliberation, 'needs to be institutionalized in the public sphere through legal and juridical means' (Huxley 2000: 371). Moreover, collaborative planning cannot do without planning law and property rights, for the 'collaborators' will want to see their interests (the outcome of the collaboration) secured by the formal assignment of new property rights, as for example in a formal plan.

Conclusion

We can conclude, that planning by law and property rights has its strengths, namely it creates stability and security. A weakness emerges from this strength: it may create lock-in situations which are difficult or expensive to 'unlock'. Collaborative planning – encompassing other modes of spatial governance than planning by laws and property rights – can help to reduce inflexibility by giving more room to market actors and thus creating flexible modes of governance, and it can help to tackle wicked planning problems by a more deliberative planning. Nevertheless, planning by law and property rights is indispensable whenever a public agency wants to carry out spatial planning. We need to know under what conditions its weaknesses are important. Under those conditions, it can perhaps be complemented with collaborative planning.

Collaborative planning, however, also has its weaknesses, regarding the democratic legitimacy or the wishful thinking in consensus-oriented planning. So, collaborative planning approaches always need to be carried out with planning by law and property rights as a complement: planning law determines the boundaries and frontiers of collaborative planning (Stierand 1993). The weaknesses of collaborative planning too should be recognized and tackled.

A possible combination between regulative and collaborative planning, a combination which should be adapted to the circumstances, is the task to be worked on when theorizing planning by law and property rights.

Overview of the Book

We have divided the book into four parts to reconsider planning by law and property rights. In the first part, we discuss the methodological aspects of such planning. In the following three parts, we discuss how to cope with the institutional, environmental, and economic (or financial) constraints on planning by law and property rights.

Methods for Exploring the Possibilities of Planning By Law and Property Rights

The contributions by Needham, and by Geuting, and Needham, put forward methods which can be used to explore what can be achieved by this sort of planning, and how the effects can be better predicted.

Needham proposes thought experiments. By looking at what is the essence of planning by law and property rights, it is possible to think about the two approaches to achieve a desired planning outcome: by public regulations about how property rights may be used (this is the usual planning approach), and by structuring markets in property rights. What is the relationship between both approaches, and under what circumstances might one be preferred above the other? This can be explored using 'thought experiments'.

Geuting and Needham consider the possibilities of planning by creating new property rights and structuring markets in them. How can we do that, without actually changing the law? Geuting and Needham propose using thought experiments and game simulation. They describe how game simulation has been used as a method to experiment with law and property rights under real life conditions. Their contribution uncovers and discusses the difficulties with this, and encourages others to develop such methods further.

Coping with the Wider Institutional Framework

Planning by law and property rights is embedded in the wider institutional context of the society, and this restricts the possibilities of changing planning law and property rights purely in order to achieve planning aims. The contributions in the second part of the book explore in which way the institutional framework constrains planning by law and property rights and the consequences for attempts to change planning law.

Oranje and Berrisford explain how, when apartheid was banished in South Africa, hopes were high for creating a new society, and spatial planning was to contribute towards this. However, they also show that attempts to introduce new planning laws met with great frustrations. It was not so much that political reactionaries tried to prevent the loss of the old situation, rather that planning law has to be embedded in a wider institutional framework, and that framework – containing elements from both the ancient regime and the new republic – reflected the wishes and interests of so many different parties (many of them governmental), that the new planning law desired by the reforming planners was thwarted and stunted. The text also requires that we think soberly about what we can expect from spatial planning. Needham states that land-use planning 'can make a person richer or poorer' (Needham 2006: 3); taking that statement to extremes could lead to the assumption that, applied in a certain way, planning can initiate societal changes and discriminate between certain groups in a society. Oranje and Berrisford describe as 'overly-optimistic' the expectation that planning law can lead to 'an improvement in the lives and life chances of the deprived African majority'. Spatial planning is indeed in the powerful position of being able to influence stakeholders' decisions about the allocation and distribution of resources – in particular of land. But to what extent can planners use it to create socially fair results? The contribution by Oranje and Berrisford helps to identify the institutional boundaries of planning by law and property rights. Restrictions in South Africa result from institutional settings – 'the minefield of government institutions' (see Oranje and Berrisford, Chapter 3) – and intergovernmental dynamics, and from the tensions 'between the agents involved' (op.cit.). With that latter, the authors say that inter-subjective aspects also play an important role, an aspect that, particularly in the field of planning by law and property rights, is often neglected.

It is curious that recent literature on planning theory pays so little attention to planning by law and property rights. An exception is Jacobs who shows in his contribution how discourses about the extent of governmental influence on property rights are conducted in the US and Europe. In the US, the rise of the conservative Tea Party is challenging the discourse followed by most established planners. And in the Europe of the EU, conservative ideas about property rights are gaining ground. Such changes in the debate can have great consequences for spatial planning, since they question and discuss the justification of (governmental) planning interventions in property rights. Jacobs also reflects on the scope, frontiers, and boundaries of spatial planning.

Hartmann and Spit discuss a particular institutional entrenchment when they present the struggle of German water managers to apply a new European regulation. Germany is faced with the need for better adjustment between water management and spatial planning. The European Flood Risk Management Directive asks for a new mode of water governance. This mode is unusual for German water managers; but also for spatial planners, who do not expect water managers to act out this mode. This leads to unexpected problems with implementation, for modes of governance cannot be just legally imposed. Hartmann and Spit discuss one of the reasons, namely that the legal structure of the laws for water management is different from that of the laws for spatial management. The first uses condition-based laws, the second more performance-based laws. This becomes clearer when the adjustment in Germany is compared with the adjustment in the Netherlands.

Coping with Environmental Change

Changes in the environment (such as climate change, water stress, natural disasters) can lead to situations, where land uses need to be changed or adapted. Inflexibilities and lock-in situations can become entrenched due to unforeseen environmental changes. The following chapters discuss environmental change and its implications for planning by law and property rights. Planning in a time of global environmental change requires making decisions now which take account of uncertain environmental conditions many years in the future. These contributions explore the role of planning by law and property rights under those circumstances. Tennekes, but also Wegener, and Tarlock look at spatial planning as an institutionalizing of environmental constraints. Seen in this light, planning needs to cope with inflexibility and lock-in situations caused by unforeseen environmental changes. Such constraints fundamentally frame the scope of planners. Natural constraints determine the scope for planning decisions. In a changing environment, planning decisions might need to be revised or abandoned. Then, inflexibility and lock-in situations can not only be expensive (for example the costs in the aftermath of a natural disaster), but also dangerous.

Planning decisions taken now about the physical environment can have consequences many years into the future. But environmental changes might be such that those decisions turn out to be wrong or unfortunate. Tennekes describes spatial planning as an exercise between accepting and adapting to environmental forces that constrain human activity, and on the other hand manipulating them (predominantly with methods of modern engineering). The Netherlands for example is exploring at the moment options for building floating cities as an adaptation to climate change. Such changes require adaptive and flexible planning schemes. The Netherlands in particular can look back to a successful history of manipulating environmental constraints (or mitigating negative environmental impacts), predominantly by using engineering methods to cope with water related issues. But this mitigation has its boundaries, even in the Netherlands. How can planning decisions take that into account? Tennekes proposes adaptive planning, shaping the conditions for future planning decisions in such a way that they will take into account the best knowledge then available on changing climatic conditions. What institutional structures are best for such adaptive planning? They must be formal, for legal regulations are necessary to regulate matters over such a long time. And many of those regulations will concern formal planning law and property rights.

Tarlock looks at environmental changes from the perspective of property rights. Property rights enjoyed now by an owner might prevent changes in the future which are necessary to adapt to climate change. In addition, Tarlock illustrates very clearly how climate change introduces new risks for landowners. For that reason, it would be socially useful to create property rights which adapt to changed environmental conditions. Tarlock questions whether property rights are static – as many libertarians would have us assume (Coase, 1960). Such an idea is not as radical as might appear, for even now some property rights have that inbuilt flexibility. Seen in the light of environmental policy, Tarlock's proposition can be framed as follows: Property rights encompass a right and a duty: the owner has the right to enjoy the exercise over her property, other have the duty not to interfere with the owner's exercise of that right (Cooter and Ulen, 2004: 77–78). One important function of the state is to protect a person's property right by regulating land uses so that interferences are minimal, and to penalize those who interfere with someone's land use. However, this requires a state that is able to identify the interferer and which has the power to impose restrictions on him or her. This is the polluter-pays-principle. The alternative to this is, besides the precautionary principle, the principle of common burdens (Hartkopf and Bohne, 1983: 86). The polluter-pays-principle regulates and set incentives to avoid interferences. The principle of the common burden is used in cases where a polluter cannot be identified. The latter can be the case in exceptional environmental disasters. If – due to environmental changes – the risk of such environmental disasters increases, it burdens the state. Tarlock argues for property rights which restrict landowners from putting themselves in a risky situation (where either the state must compensate, or there is no compensation). This would be – in terms of environmental policy – an application of the precautionary principle to property rights. Seen in this light, Tarlock's proposition seems not too radical, but a logical consequence of environmental changes.

Wegener states that planning for sustainable development requires that the effects of current planning decisions and of environmental changes in the far future need to be taken into account now. This, he concludes, cannot be done if the planning decisions are taken at a governmental scale which is too low, nor by collaborative planning which gives much weight to commercial parties which inevitably have short term interests. He argues that planning institutions need to be assigned so that particular self-interests of stakeholders do not prevent them from making objective planning decisions. For Wegener, planning by law and property rights crucially depends on planning levels. Coping with environmental challenges of the future, he argues, requires a reconsideration of the principle of subsidiarity.

Wegener, Tennekes, and Tarlock show in their contributions that – no matter whether climate change is man-made or just a normal variation in the long-term world climate – planners need to take the changes into account, because mitigation alone is not capable of tackling the changes; adaptation is required. This means that planners need to find legal arrangements to cope with environmental challenges. These legal arrangements are social constructions of environmental constraints.

Coping with Uncertainty and Lock-ins

Planning by law and property rights can create certainty and thus help to stabilize land and property markets. But only if the planning agency can forecast well. There is no guarantee of this. And what happens when planning forecasts are wrong, and when, moreover, the planning is carried out not just by law and by property rights, but also by active public involvement in property markets? If it is wished to change the plans, this might destroy those property values. The resulting mistakes are 'locked-in' because of the new property rights that have been created. So what kinds of plans should be made? The next three chapters discuss these issues.

Ploegmakers and van der Krabben start from the argument that planning by law and property rights can reduce the uncertainty faced by the users and creators of the physical environment. One way of doing this is by forecasting the demand for spaces and buildings and by making spatial plans accordingly. But how reliably do planning agencies do that? If that is poor – and the authors show that it is, when Dutch municipalities forecast the demand for employment land – the spatial plans could produce an end result which is locked in and difficult to change.

Janssen-Jansen shows another example where spatial plans forecast the future demand incorrectly. The offices in the Amsterdam region are an example of oversupply caused by past planning decisions. The chances of that happening are greater when the planning agency has a financial interest in allowing development, and when commercial parties are directly involved in making the content of the plans, for they too want profitable development to be allowed. Appropriate governmental structures could prevent this, such as intervention by higher level authorities to rein in municipalities. When, for such reasons, far too many development possibilities are included in spatial plans, a lock-in is created which, because the new property rights have a high value, can be very difficult to resolve. By the time that oversupply becomes apparent, it can be very expensive to tackle it by withdrawing development rights.

Buitelaar finally points out that plans might turn out to be undesirably inflexible. They might prevent unforeseen developments which are later considered to be desirable, or they might have to be changed at great cost. One way of avoiding this would be to make the plans which look into the future indicative, and to make legally binding plans only when it is desired to conserve a particular area in order to prevent it from change, or to regulate the legal acceptance of a proposed development which has been found to be desirable. An empirical analysis of the types of plans made by Dutch municipalities shows that they do, in fact, do that.

The chapters by Janssen-Jansen, Ploegmakers and van der Krabben, and Buitelaar show the dilemma of planning which is very growth-oriented. It can end up in lock-in situations, if the growth turns out differently than expected. The case of the office tragedy in Amsterdam illustrates that problems caused by inflexibility can be more difficult to resolve in circumstances of economic decline than economic growth. Few planning laws include the provision for 'down-zoning' building zones to, for example, 'no development', in any case not without huge compensations, as Davy found for the German planning system (Davy, 2006).

The book is rounded off with a chapter by Hartmann and Needham, in which they draw conclusions from the ensemble of the various chapters. This sheds light on different aspects of spatial planning and from different perspectives. Planning by law and property rights is reconsidered in many of its facets, so that it can be used better in the face of contemporary challenges to spatial planning. Perhaps the most important message in this book is that we cannot neglect and should not reject planning by law and property rights, but continually reflect on it. This book contributes to such reflection.

References

Albrechts, L. 2004. Strategic (spatial) planning re-examined. *Environment and Planning B: Planning and Design*, 31(5), 743–758.

Allmendinger, P. 2002. Towards a post-positivist typology of planning theory. *Planning Theory*, 1(1), 77–99.

Allmendinger, P. and Haughton, G. 2010. Spatial planning, devolution, and new planning spaces. *Environment and Planning C: Government and Policy*, 28, 803–818.

Arthur, W.B. 1989. Competing technologies increasing returns and lock-in by historical events. *The Economic Journal*, 99, 116–131.

Baum, H.S. 1977. Toward a post-industrial planning theory. *Policy Sciences*, 8, 401–421.

Bertolini, L. 2010. Complex systems, evolutionary planning?, in G. de Roo and E.A. Silva (eds), *A Planner's Encounter with Complexity*. Farnham: Ashgate, 81–98.

Billé, R. 2008. Integrated coastal zone management: Four entrenched illusions. *Surveys and Perspectives Integrating Environment and Society*, 1, 75–86.

Bromley, D.W. 2000. Regulatory takings and land use conflicts, in M.D. Kaplowitz (ed.), *Property Rights, Economics, and the Environment*. Stamford, CT: JAI Press, 23–33.

Coase, R. 1960. The problem of social cost. *Journal of Law and Economics*, 3, 1–44.

Cooter, R., and Ulen, T. 2004. *Law and Economics* (4th edn). Boston, MA: Addison-Wesley.

Davy, B. 1997. *Essential Injustice: When Legal Institutions Cannot Resolve Environmental and Land Use Disputes*. Wien, New York: Springer.
Davy, B. 2005. Bodenpolitik, in E.-H. Ritter (ed.), *Handwörterbuch der Raumordnung*. Hannover: ARL, 117–130.
Davy, B. 2009. Land thrift as sustainable development: the case of Germany's 30 hectares goal, in A. van der Valk and T. van Dijk (eds), *Regional planning for open space*. London, New York: Routledge, 279–300.
Davy, B. 2010. Freiraumsicherung durch Bodenpolitik: Was passieren müßte wenn wir das Ziel-30-ha ernst nähmen, in K. Marion and K. Selle (eds), *Siedlungsflächen entwickeln. Akteure. Interdependenzen. Optionen*. Detmold: Rohn Verlag, 258–271.
Durner, W. and Ludwig, L. 2008. Paradigmenwechsel in der europäischen Umweltrechtsetzung? *Natur und Recht*, 30(7), 457–467.
Ellickson, R. 1993. Property in land. *Yale Law Review*, 102, 1315–1400.
Forester, J. 2004. Reflections on trying to teach planning theory. *Planning Theory and Practice*, 5(2), 242–251.
Frank, S. 2006. The European Union and the European cities: three phases of the European urban policy, in U. Altrock (ed.), *Spatial Planning and Urban Development in the New EU Member States. From Adjustment to Reinvention.* Aldershot: Ashgate, 39–56.
Fraser, N. 1990. Rethinking the public sphere: a contribution to the critique of actually existing democracy. *Social Text*, 25/26, 56–80.
Goodin, R.E. and Tilly, C. 2008. *The Oxford Handbook of Contextual Political Analysis*. Oxford: Oxford University Press.
Healey, P. 2003. Collaborative planning in perspective. *Planning Theory*, 2(2), 101–123.
Healey, P. 2009. In search of the 'strategic' in spatial strategy making. *Planning Theory and Practice*, 10(4), 439–457.
Heinz, W. (ed.) 1993. *Public Private Partnership — Ein neuer Weg zur Stadtentwicklung?* Stuttgart, Berlin, Köln: Kohlhammer.
Hillier, J. 2010. Introduction: Planning at yet another crossroads?, in J. Hillier and P. Healey (eds), *The Ashgate Research Companion to Planning Theory. Conceptual Challenges for Spatial Planning*. Farnham: Ashgate, 1–34.
Huxley, M. 2000. The limits to communicative planning. *Journal of Planning Education and Research*, 19(4), 369–377.
Ingram, G.K., and Hong, Y.-H. 2009. *Property Rights and Land Policies*. Cambridge: Lincoln Institute of Land Policy.
Kurth, D. 1999. *Laboratorium Emscher Park*. Dortmund: Informationskreis für Raumplanung.
Needham, B. 2006. *Planning Law and Economics: The Rules we make for Using Land*. Abingdon: Routledge.
Ostrom, E. 2007. *The Challenge or Crafting Rules to Change Open Access Resources into Managed Resources*. Paper to the International Economic

Association roundtable on the Sustainability of Economic Growth, Beijing (China), July 13–14, 2007.

Paffrath, C. 2004. *Macht und Eigentum: Die Enteignungen 1945–1949 im Prozeß der deutschen Wiedervereinigung*. Köln: Böhlau Verlag

Rittel, H.W. and Webber, M.A. 1973. Dilemmas in a general theory of planning. *Policy Sciences*, (4), 155–169.

Schiller, R. 2001. *The Dynamics of Property Location München*. Sponpress: London.

Selle, K., Hüchtker, S., Scholz, B., Sinning, H., Sutter-Schurr, H., et al. 2000. *Freiraum, Siedlung, Kooperationen: Forschungsergebnisse, Hinweise für die Praxis, Folgerungen. Arbeits- und Organisationsformen für eine nachhaltige Entwicklung*. Dortmund: Dortmunder Vertrieb für Bau- und Planungsliteratur.

Stein, U. 1995. Raumplanung zwischen Staat, Markt und Gesellschaft. *Raumforschung und Raumordnung*, (5), 393–396.

Stierand, R. 1993. Neuorientierung in der Planungstheorie? *RaumPlanung*, (61), 141–147.

Thompson, M. 1997. Rewriting the precepts of policy analyses, in R.J. Ellis and M. Thompson (eds), *Culture Matters. Essays in Honor of Aaron Wildavsky*. Boulder, CO: Westview Press, 203–216.

Wegener, M. 1999. Raumplanung als Systemrationalität oder die Rettung der Raumplanung durch die Ökologie, in K.M. Schmals (ed.), *Was ist Raumplanung?* Dortmund: Dortmunder Beiträge zur Raumplanung 89, University of Dortmund, 165–72.

PART I
Methods for Exploring the Possibilities

Chapter 1

Interests and Rights in Property, and their Place in Land-use Planning: A Theoretical Investigation

Barrie Needham

Land-use planning is often seen as opposing the market and as restricting the exercise of property rights. Planning is justified by referring to the need for a public authority to act in the public interest. This chapter rejects that line of reasoning. Instead, it focuses on the interests which people have in the way in which land is used. Some of those interests are protected by private law: they are property rights. The holders of property rights can exercise them in the market. Other private interests in land are recognized by the state, but are not given the status of property rights. Those interests are protected by a public authority, using public law, on behalf of its citizens. This is the activity of land-use planning. The resulting land use is the result of people exercising their property rights in a legal context imposed by a public authority in order to protect the private interests of others in land. However, the distinction between interests which are protected as property rights and interests which are protected by land-use planning is not absolute and is the result of a political decision. Using these ideas, this chapter investigates whether some of the goals of land-use planning could be realized by changing interests into rights, in such a way that markets in those rights would realize the desired land use. This chapter goes no further than an investigation of feasibility: it does not recommend replacing land-use planning by markets in property rights.

I start with a working definition of land-use planning. This refers to the activities of public authorities when they try to achieve some politically determined goal by deliberately influencing the way in which land in a certain area is used. The goals aimed at in this way can be various: more efficient production, reducing unemployment, improving conditions in residential areas, reducing accidents from traffic, transport which is more ecologically sustainable, protecting wildlife habitats, making town centres more pleasant, providing more affordable housing, and so on. A public authority expects that if the desired land use is realized, the goals will be (more nearly) reached. So it carries out land-use planning in order to realize that land use.

In this chapter I question neither the legitimacy nor the value of the state trying to influence land use on behalf of its citizens. Nor do I question whether the assumed causal link between the desired land use and achieving the goal is always correct. But I do question the methods by which the public authority carrying out that planning tries to realize the desired land use. The chapter is based on Needham (2006), where the argument is worked out in much more detail, and is a reworking of Needham (2005b).

The usual way of carrying out land-use planning is by making building and development conditional on the granting of a permission, which is given only if the initiative conforms to a plan or policy. We can call this 'regulation'. However, there is a possible alternative to this, namely creating markets in property rights in such a way that the initiatives which people take (automatically) realize the goals of the planning policy. To the extent that this alternative is feasible, it has possible advantages over the usual way of carrying out land-use planning, both for the citizen and for the planning agency. People are freer in how they decide to use their land and buildings. Also, if people enjoy property rights, they can use them to protect themselves against those who have excessive power. At the same time, people have more responsibility for creating and maintaining their own surroundings. For the planning agency, it would have advantages of less direct involvement, less financial risk, and less risk of freezing land use in a state which turns out to be inappropriate. The question is: is it technically possible to arrange this in such a way that the publicly chosen goals of land-use planning are still met?

To investigate this, we start by looking at how the state can influence the way in which land is used.

Influencing Land Use

Achieving the goals of land-use planning depends on how land is *used*, not on who *owns* it. Moreover, under land-use planning, the decisions which people make about using land, although they are restricted, are usually taken voluntarily. In most cases, there is no authority which says how the owner of a parcel of land or building *must* use it, nor that the owner may not buy or sell, nor who may own what land and who may not: it is not 'command and control', even though that term is often applied to it. A public authority which wants to influence the way in which land in a given location is used, needs therefore to understand how the owners make decisions about using their land. The public authority itself has no right to take those decisions, unless it owns the land. Land-use planning will succeed only if it can *influence* the private decisions.

How can that influence be exerted? The way we are most familiar with is development control. This is how it is described in English planning terms, in other countries it is called the building permit, or the zoning permit, or some such. The content is always more or less the same: if you want to change the use of your land, you must first obtain permission, which is granted only if the proposed change conforms to the land use desired by the public authority. Other ways which can be found in practice are: giving a subsidy to a land-use decision which is desired, or imposing a levy on an undesired land-use decision; persuasion and encouragement to take the land-use decision which the public authority desires; and providing roads and other infrastructure which makes some uses more attractive than others. The decision remains free, but is influenced or constrained. The decision ceases to be free if ownership is taken away, amicably or compulsorily. That is common when no-one besides the public authority is interested in making the desired land-use change, for example providing open space or a road.

Land-use Planning and Interests in Land

Why should a public authority want to influence decisions about land use? The answer is that the land use resulting from all the decisions made by private (legal) persons is not always what the public authority wants. So that authority wants to change those decisions 'in the public interest'. This formulation I do not find satisfactory. It implies that in a particular area one way of using land is more 'in the public interest' than another way. That is exceedingly difficult to substantiate (see, for example, Lucy 1988).[1] Economics offers the substantiation in terms of 'distortions' caused by external effects, monopoly conditions, public costs, and so on: but those arguments too no longer convince (see for example Lai 1994; Needham, Segeren, Buitelaar 2011).

One of the main problems with the idea of 'the public interest' is that it suggests that you and I, as individuals, might have to be constrained or influenced in the interest of 'the public'. But we too are the public! And who are the other members of the public who will be benefited by constraining us? Talking of 'the public interest' has a similar effect to talking – as many economists do – of 'social costs' (Bromley 1991: 19). The others, who are affected by our actions, are de-personalized into an amorphous and unidentifiable 'public' or 'society'.

In order to avoid that depersonalisation, we need to look more closely at peoples' interests in the way in which land is used. It is clear that, if I own a parcel of land, I have an interest in the way in which that parcel is used. But I have an interest too in the way in which my neighbour uses her land, for that can affect me. For example, I do not want her to dig such a deep hole in her garden that the foundations of my house are endangered. And on a larger geographical scale, and whether or not I own land, I have an interest in how others use their land. I have an interest in my neighbour not painting her front door bright pink. I have an interest in how much traffic uses my street. I have an interest in the vitality of the shops in my town centre. I have an interest in the conservation of an open landscape. I have an interest in clean air.

1 Following Alexander 1992 we make a distinction between a substantive and a procedural interpretation of the public interest. Saying that a particular land use is in the public interest is an example of the substantive interpretation.

We can now give a different answer to the question: why does a public authority want to influence the land-use decisions made by private persons? It is because the decision by one person about the use of her land can affect the interests of another person. The authority is acting not so much to uphold some general public interest, as to uphold the identifiable interests of many identifiable private persons.

Land-use Planning and Rights in Land

We can conclude that public authorities practise land-use planning in order to protect the interests of people in how land is used. However, the state protects the interests of private persons in land in another way too, namely by creating and protecting property rights. And – it will be clear – those two ways can clash, and often do clash. The right to own landed property includes the right to use that property in a particular way. Moreover, the right of ownership is strongly protected (see for example article 1 of the Protocol to the European Convention on Human Rights: 'Every natural or legal person is entitled to the peaceful enjoyment of his possessions'). Land-use planning can restrict the exercise of that right. So it is not surprising that land-use planning is often politically contested. Before we investigate further the relationship between a public authority upholding the interests of its citizens by land-use planning, and the state upholding such interests by making of them property rights, it is useful to put the right of ownership in perspective.

Most countries have an extensive legal system of property rights regulating the use of land. The reason is that land use is very important for the functioning of the society, both economically and socially. Property rights give a person security to use a piece of land in a particular way, and that security helps the market to work efficiently and stabilizes social relationships. Landlord/tenant legislation illustrates that well. The landlord has certain rights, the tenant also, and those rights help the market in tenancies to work. Ownership is, however, never absolute (in spite of the statement in article 17 to the 'Declaration des droits de l'homme en du citoyen' in 1798 that ownership was 'inviolable et sacré'). The most important restriction is that the owner is not allowed to use the land in a way which harms others or which restricts others greatly. The dog is not allowed to sleep in the manger, even if he owns it, if the horse has no other source of food. Squatters can claim rights from landowners. How the law regulates this varies greatly from country to country, but the principle is the same everywhere (Honoré 1961). It is the details which are interesting. For in spite of the general principle, most countries *do* allow the owner of land to use it in ways that harm others, if that use gives more benefit than the harm it causes (Coase 1960).

There is a second type of restriction too, namely when the owner of the freehold right voluntarily restricts herself by granting a partial right to someone else. That other person acquires the right to use the land in a particular way, such as a ground lease: and the court protects the holder of that partial right, if necessary against the owner of the freehold right.

Those two types of restriction have nothing to do with land-use planning. They concern the 'traffic' between private legal persons. If one private person harms another, or interferes with the exercise of a right by another, that other (the plaintiff) may take the doer of the deed (the defendant) to court. The action is taken by a *private* person. The court, of course, has to be a *public* institution, for it might be necessary to use force against one of the two parties, and the state is the only agency legitimized to use force. But, crucially, the initiative to enforce the restriction is taken by a private person, not by the state. However, the state will not uphold laws which it does not support. It follows that the state must define property rights (or, under case law, it recognizes property rights): what they allow and do not allow, how they will be enforced, the punishment for transgressing a right, and so on.

In order to understand the full significance of this, we have to refine our understanding of a property right. It is a way of using land (and buildings), *where that way is recognized by, and protected by, the courts.* That addition is of paramount importance. If, for example, I rent my land to a farmer, I give that person rights over my land. If I want to get rid of the farmer in order to sell the land to a housing developer, the courts will not recognize my action if the right which the farmer has acquired from me protects him from eviction. This fact undermines yet further the idea that property rights are or can be absolute, or that they are in some way indisputable, natural or inalienable (as propounded by, for example, Locke in his 'Two treatises of government', 1689), for they are created by the state. 'It is the fact that something is protected that makes it a right, rather than that something is protected because it is a right' (Bromley 1998). It follows that different state governments can create different property rights. And those international differences exist in practice, as empirical investigation shows, and as the examples later in this chapter illustrate.

Two Ways in which the State Protects Interests in Land

There are, therefore, two ways in which the state protects peoples' interests in land. One way is to declare an interest a property right. That declaration is accompanied by rules allowing the owner of a property right to protect her right in the courts. That we call the 'private law' way of protecting interests: the holder of the interest uses private or civil law to protect herself. The other way is when the state gives a public authority (such as a municipality) powers to influence the way people with property rights exercise their right to change the use of the land, where the public authority does that in order to protect the interests of those who might be harmed by the exercise of property rights by others. This we call the 'public law' way of protecting interests: the public authority uses public law – taking to court someone who breaches the rules, such as building contrary to a building permit – in order to protect the interests of those who cannot protect their interests in any other way. Clearly, when the state creates those two possibilities – by private law and by public law – it has to ensure that they are consistent. In particular, if the state allows someone's property rights to be 'taken' by public law, it might be necessary to compensate the holder of that right.

The question then arises: which of the many interests that a person can have in land does the state protect by giving that person a property right, which of the interests are protected on behalf of the person by a public body authorized to do that, and which are not protected at all?

The answer to this is partly political. I am given no right to protect my interest in the colour of my neighbour's front door, nor will a public authority do that for me. The reason is that, it has been decided, *my* interest in that colour is less important than *her* interest in the colour.

The answer is partly technical too. If my neighbour damages the foundations of my house, there is a clear connection between the cause and the effect, and between the person responsible for the cause and the person harmed, moreover there are just a few people involved. So it is feasible for me to exercise a right to take the perpetrator to court for the damage caused to me. And, in fact, in many countries I do have that right (the 'right of support'). If, on the other hand, a factory pollutes the air breathed by hundreds of people (the effects are diffuse), that connection between cause and effect is less clear; moreover it would be extremely difficult for those hundreds of people to co-operate to take an action against the factory. So giving a person a property right to clean air would be ineffective in protecting that person's interest in clean air.

It is for that reason that my interest in the foundations of my house not being damaged is protected by giving me a property right, but my interest in clean air not. Nevertheless, the state recognizes that latter interest. And it protects it by *public measures* such as conditional permissions (a building permission will not be given if the factory is too near to housing) or an outright prohibition (limits on the emission of certain gases).

However, the decision to protect some interests by private law and other interests by public law can always be revised. This is often not recognized, for we have the tendency to think of property rights (the private-law rules) as being absolute. Enough has been said above to show that property rights are never absolute, also that they are not 'natural' but created, or recognized, by the state. It follows from this that property rights can be a 'policy variable'. A well-known example is inheritance laws, which are designed to prevent an unhealthy accumulation of wealth. Policy rights can be purposefully changed in order to achieve a particular outcome. It must be added that rights in land often have a high ideological content, as a result of which they are not changed easily or quickly. Moreover, international agreements, such as the European Convention on Human Rights, might restrict a national government from making some changes to property rights.

The corollary is that, as property rights influence the way land is used, and as property rights have been created, then property rights can be deliberately changed in order purposefully to influence land use. We are familiar with such policy, under the heading of 'land tenure reforms'. But we do not usually make the link with land-use planning, nor recognize that creating and changing property rights can be used as an instrument of land-use planning. Here we make explicit, that it can be decided to pursue some of the goals of land-use planning by changing property rights, thus shifting from public-law rules to private-law rules.

We work out this idea below. First we illustrate it briefly by applying it to the wish that non-replaceable resources be used sustainably. Usually that goal is pursued by restrictions (public law) on using such resources. If we look at the interests involved, we see that by pursuing our interests *now*, we are endangering the interests of *future generations*. Perhaps we can protect those future interests by making them into property rights (Bromley 1991: 19). Those rights would then be vested in the generation which will be 21 in – say – 50 years, and trustees would be appointed to protect those rights until then.

Land-use Planning and Markets

It is now time to introduce the concept of 'the market'. We started by pointing out that most decisions to change the use of land are made voluntarily. We can express this by saying that most of those decisions come about 'in the market'. This might be a broader way of using that term than some are accustomed to. There is a market when three conditions are met. One person has something which another person wants; an exchange takes place voluntarily, whereby the 'demander' is free in the choice of 'supplier' and the 'supplier' is free in the choice of to whom she supplies; and that exchange takes place regularly between more people (Hodgson 2008). Lindblom (2001) points out that it is almost impossible to conceive of processes, such as those by which land use comes about, where there are no voluntary decisions. That is, processes where the market is excluded.

It follows from this, that a discussion in terms of 'planning or the market?' is not useful. The market cannot be excluded from influencing land use. The only question which is interesting is about the *mix* of planning and the market (Buitelaar 2003). It follows also, that when someone exercises her property right in land, that is an action 'in the market', for property rights set a framework for exchanging interests in land. So the decision whether an interest in land should be protected by private law (property rights) or by public law (planning by regulation) is a decision about the mix of planning and the market. Another conclusion is that the decision to shift that protection from public law to private law by creating or changing property rights is tantamount to creating markets in property rights. For that shift creates new exchange structures, within which people can interact freely and willingly (see Chapter 2).

Creating Markets in Property Rights

Sometimes there can be a real choice for the state between protecting a person's interest on the one hand by creating an appropriate property right and on the other hand by imposing public measures. This choice has been the subject of a number of publications. Coase (1960) examined the economic arguments, by pointing out the reciprocity of the actions: if one person's actions damage me, then restricting those actions benefits me but harms the other person. 'The question is commonly thought of as one in which A inflicts harm on B and what has to be decided is, How should we restrain A? But this is wrong. We are dealing with a problem of a reciprocal nature. To avoid the harm to B would be to inflict harm on A. The real question that has to be decided is, Should A be allowed to harm B or should B be allowed to harm A?' (op. cit.: 96). So the decision to elevate an interest into a right should be based on a study of the economic effects of making the interest legally defensible. Ellickson (1973) made a most thorough investigation with the title 'Alternatives to zoning: covenants, nuisance rules, and fines as land use controls'. He was motivated by the preference for private law measures (that is, through property rights and tort law) over public law measures (such as zoning). He concluded that private law measures could effectively be used in some situations, but not in all. And more recently, Webster and Lai (2003) have conducted a similar exercise, to investigate whether problems caused by 'resources in the public domain', which are usually tackled by public law measures, could not better be tackled by changing property rights so as to make those resources 'club goods'. Those investigations must not be confused with attacks on land-use planning by those who resist any restriction on the exercise of individual property rights. For those latter attacks either deny the external effects of the private use of property (that is, they deny that there is any 'public interest' involved), or they deny the ability of a public authority to protect the interests of 'the public' (A classical statement of this 'libertarian' stance is to be found in Sorensen 1981).

The investigations I have mentioned do not reject the goals aimed at by much land-use planning, nor the legitimacy of the state influencing land use. Rather, they ask the question: is it always necessary to try to achieve those goals by public law regulation of the use of property rights? Could they not perhaps also be achieved by letting people use their property rights? And – and this is crucial – the question recognizes the possibility of changing existing property rights and creating new ones. The question posed in this chapter is: would it be technically possible to achieve land-use goals in this way?

To answer this, we have to take a broader approach than some economists do. These sometimes assume that the goal of all public policy is maximum economic efficiency, and that all varieties of public policy (such as for land use, but also for health, for nature protection, for education) must in the end strive to achieve that. I do not make that assumption. I take the approach described by Williamson (1995, 1999) as 'remediability', and I evaluate different ways of achieving the same policy goals by comparing the effectiveness with which they do that. So the question becomes: which of the goals commonly adopted for land-use planning could be achieved by creating new markets in property rights and by withdrawing some of the public controls on land use? Would those goals be achieved better (more effectively) in that way? What would be the consequences for the use of scarce resources? And what would be the distributional effects? It will be clear that we cannot answer all those questions in one short chapter. Answers are given in Needham (2006) and Needham (2005a), where also the theoretical arguments are given much more fully. Here I do no more than report briefly a few of the conclusions of those investigations, using four types of situation where deliberate use of property rights might be an alternative to customary land-use planning.

Amenity in Housing Estates

It is an aim of land-use planning that housing estates be developed where residents can live in pleasant surroundings and with the knowledge that those surroundings will continue to be pleasant. One way of doing that is by making a land-use plan which specifies the conditions which must be met by any development on the estate, by refusing a building permit for an application which does not conform to those conditions, and by monitoring subsequent changes to check that they do not transgress the conditions. This might be coupled with an environmental policy, also realized by granting or refusing permits.

An alternative way of achieving the same goals is by making it possible for all people who buy a house in a particular area to enter into a covenant, whereby they restrict themselves from taking certain actions which would damage the amenity of the estate, and by vesting the right of covenanter in a residents' association. This association is then responsible on behalf of its residents for monitoring whether a house owner fulfils the terms of the covenant and, if she does not, for taking legal action on behalf of the other residents. The *interest* which the residents have in the amenity of their area has been transformed into a *property right* (Grant 1988). That might seem to be a draconian measure: but it is entered into freely by very many people in the United States ('private community associations') and also, but less commonly, in England ('a scheme of development'). Clearly, there are many people who are prepared to restrict themselves, as long as their neighbours also are similarly restricted, in order to protect the amenity of their housing estate. And equally clearly, this is a feasible alternative to pursuing the same goal through land-use plans and building permits.

It illuminates my general argument to ask the question: why do we not find similar arrangements in the Netherlands? The answer is: such an arrangement is not legally binding in that country, for the system of property rights there does not permit people to be obliged to become a member of a residents' association (Van Velten 2003). In other words, Dutch property rights do not allow a market solution to the problem of achieving and maintaining residential amenity. Dutch property rights could be changed so as to make that possible. That is an example of how a market in property rights could be created, thus providing an alternative way of pursuing some of the goals of land-use planning

Well-maintained Industrial Estates

It is an aim of land-use planning to make it possible to develop industrial estates, which have the properties wanted by the firms operating there, and where those properties are protected and maintained. The way in which land-use planning can do this is similar to the way in which that planning creates housing estates with the required amenity: building/planning permits supplemented with environmental permits. This is, however, not effective in *maintaining* the quality of the estate.

In England and the United States, this shortcoming can be overcome if someone has a right in land which makes it possible and profitable to maintain the desired qualities. The solution is easy and well known. A developer constructs an estate, sells the completed estate to a property investor, which then leases the buildings to the firms which occupy them. The investor then maintains the estate as a whole. If he does this well, his investment will retain its value. The quality is maintained by a 'market instrument'. The individual firms have a lasting interest in the quality of their estate, but it is not feasible to change that into individual property rights which could be used for that purpose. So the interests of all the separate firms are taken over by one landlord, who does have the property right which is necessary to protect that interest.

Once again, this solution is not possible in the Netherlands, because the system of property rights does not permit it. If the firms on the estate own their buildings, it is difficult to oblige them to become a member of an association which maintains the shared spaces. Moreover, the firms are reluctant to rent their buildings for a long period, because if they want to move before the end of the tenancy agreement, the remaining term cannot be sold to the next tenant (the tenancy is not tradable): so there is little *investment* interest in industrial buildings. The result is that, although many firms have an interest in maintaining the quality of the estate, no-one has property rights which would enable that to be done. As a result, the estate deteriorates quickly, and the only way of improving it is by the expenditure of huge amounts of public money. The market solution could be introduced into the Netherlands by changing a property right, namely by making it possible to sell unexpired rental leases.

Urban Public Spaces

In towns there are many open spaces which are freely accessible to everyone – urban public spaces. These are parks, shopping centres, squares, etc. They are usually provided and maintained by a public body, precisely so that everyone may use them. They are important, however, not only to the public but also to the users and owners of property adjoining or near to those spaces – dwellings, offices, shops. Those latter property users and owners have an interest in the public spaces, because those spaces give satisfaction to the users and because they affect the value of the properties. But when the spaces are provided and maintained by the municipality, those property users and owners have few possibilities for influencing the form of the spaces, or the maintenance of them. The property users and owners have an interest but no rights in those spaces.

For that reason, urban public spaces are sometimes provided and maintained privately. Well-known examples are shopping malls. The owner allows free access to everyone, but has the right to prohibit access under certain circumstances, such as when the shops are closed, or if someone causes a public nuisance. There are also cases of (small) housing developments where the spaces between the private plots are owned and maintained by the residents themselves, but freely accessible to all, again under certain conditions (see for example Lohof and Reijndorp 2006).

In such cases, the development is given planning permission under conditions imposed by the municipality and concerning the privately owned public spaces. Those conditions specify the rights of access by the public and the duties on the owners to maintain the spaces and to guarantee the access. With such an arrangement, the property rights are different from when the urban spaces are provided and maintained publicly: property rights have been created deliberately so as to provide a certain type of urban environment (Needham 2007b).

Coherent Regional Development

The aim of most land-use planning at the scale of the region is to take account of the interactions between the various activities in different locations within the region. If all the workers live in one location and all the workplaces are in another, this will result in many long work journeys, which are undesirable because of the amount of time lost, the congestion, the transport infrastructure, the traffic fumes, etc. If attractive countryside and valuable farmland are being built upon, while there is abandoned and developable land elsewhere, this is undesirable because of the loss of amenity in one location and the under-use in another. If houses are built in much dispersed locations and at low densities, the experience of open countryside that many people value is reduced and public transport to serve those houses is uneconomical.

The usual planning (public law) way of influencing land-use decisions to take account of such interactions is with land-use plans which allow certain types of development in certain locations and forbid it in other locations. Often this is supplemented with the provision of roads etc. which influence decisions in a complementary way. Could that be replaced by creating markets in property rights, so that people automatically take account of the interactions?

It is clear that this is much more difficult than at the scale of the housing and the industrial estate. The reason is that the interests which are affected by the land-use decisions of one person are very dispersed. There can be so many people who are affected, positively or negatively, and the effects can be so small for each of those people separately, even though the total effects are great, that it is not realistic to expect that a market in rights could be set up.

What we do find, interestingly, is that a public authority sometimes tries to achieve the desired land use on behalf of its citizens, using private law rules instead of the usual public law rules. The conservation agreements found in many parts of the United States are an example: a public or a private trust acquires amicably the right to change the use of farm land (Daniels and Bowers 1997). The farmer retains ownership rights, and has sold the development rights to an agency which does not intend to use them. That agency has acquired the rights in order to stop others using them, thus keeping the land open.

Conclusions

We have pointed out that the state provides two ways for influencing the way in which land is used: by creating property rights and enabling them to be enforced, and by land-use planning. The aim of both ways is to protect the legitimate interests of citizens when other citizens use their own land. We have investigated whether a public authority could achieve the goals of land-use planning by creating markets in property rights rather than by regulating the use of those rights. That has taken us on an interesting journey! It has thrown up all sorts of new insights which help us to understand land-use planning better. But the conclusions reached are not very radical: creating markets in property rights is a feasible alternative only under certain limited circumstances (and see Needham 2007a). Moreover, demonstrating feasibility is not the same as demonstrating desirability. This chapter has not attempted to compare the advantages and disadvantages of the two ways of influencing land use.

Nevertheless, this exploration of feasibility is a necessary preparation for evaluating how people's interests in land are protected. The usual practice of land-use planning is to give the task of protecting those interests to a public authority. If that task were performed by creating appropriate property rights – for the circumstances under which that would be feasible and effective – the results might be less inflexible than land-use planning, and would certainly avoid the public costs of compensating for 'takings'. On the other hand, facilitating developments which work by creating new property rights can be exclusive, such as gated communities, and which lead to fragmentation of the city.

Finally, I would argue that influencing land use by creating appropriate property rights and 'letting the market do its work' would be neither more politically right nor more politically left than the more usual way. Emphasizing property rights is often thought as being right wing, understandably when rich and powerful people have more property rights than poorer and less powerful people. I support the (rather old-fashioned) idea that the power of poorer people can be strengthened by giving them rights – which need not be full ownership rights – in landed property and encouraging them to use those powers. 'the Left should re-appropriate property rights' (Blomley 2004: 154) 'Property relations can be the means by which people find meaning in the world, anchor themselves to communities and contest dominant power relations' (op. cit.: 156).

References

Alexander, E.R. 1992. *Approaches to Planning*. Philadelphia: Gordon Breach Science Publishers.

Blomley, N. 2004. *Unsettling the Land: Urban Land and the Politics of Property*. New York: Routledge.

Bromley, D.W. 1991. *Environment and Economy: Property Rights and Public Policy*. Cambridge: Blackwell.

Bromley, D.W. 1998. Rousseau's revenge: the demise of the freehold estate, in H.M. Jacobs (ed.), *Who Owns America: Social Conflict over Property Rights*, Madison: University of Wisconsin Press, 19–28

Buitelaar, E. 2003. Neither market nor government: comparing the performance of user rights regimes. *Town Planning Review*, 74(3), 315–330.

Coase, R.H. 1988 [1960]. The problem of social cost, in *The Firm, The Market and the Law*. Chicago: University of Chicago Press, 95–156.

Daniels, T. and Bowers, D. 1997. *Holding our Ground, Protecting America's Farms and Farmlands*. Washington DC: Island Press.

Ellickson, R.C. 1973. Alternatives to zoning: covenants, nuisance rules, and fines as land use controls. *The University of Chicago Law Review*, 40(4), 683–781.

Geuting, E. 2011. *Marktstructurering als ruimtelijkeordeningsinstrument*, PhD thesis, Radboud University Nijmegen.

Grant, M. 1988. *Forty Years of Planning Control: The Case for the Defence, The Denman Lecture*. Cambridge: Granta Editions.

Hodgson, G.M. 2008. Markets, in S. Durlauf and L. Blume (eds), *New Palgrave Dictionary of Economics*, 2nd edn. Basingstoke: Palgrave Macmillan. Also available at www.dictionaryofeconomics.com.

Honoré, A.M. 1961. Ownership, in A.G. Guest (ed.), *Oxford Essays in Jurisprudence*. Oxford: Clarendon Press, 107–147.

Lai, L.W.C. 1994. The economics of land-use planning. *Town Planning Review*, 65(1), 77–98.

Lindblom, C.E. 2001. *The Market System*. New Haven, CT: Yale University Press.

Lohof, S. and Reijndorp, A. (eds) 2006. *Privé terrein, privaat beheerde woondomeinen in Nederland*. Rotterdam: NAi Uitgevers.

Lucy, W. 1988. APA's ethical principles include simplistic planning theories. *Journal of the American Planning Association*, 54(2), 383–397.

Needham, B. 2005a. *Een andere marktwerking*. Rotterdam: NAi uitgeverij, Den Haag: Ruimtelijk Planbureau.

Needham, B. 2005b. *Markets, Rights and Interests*. Paper presented at AESOP congress, Vienna, 13–16 July 2005.

Needham, B. 2006. *Planning, Law, and Economics*. London: Routledge.

Needham, B. 2007a. Land use planning and the law. *Planning Theory*, 6(2), 183–189.

Needham, B. 2007b. Belangen, rechten en plichten bij publieke ruimten: een andere taakverdeling?, in A. Lengkeek (ed.), *Stedelijkheid als rendement: privaat initiatief voor publieke ruimten*. Haarlem: Trancity, 194–211.

Needham, B., Segeren, A. and Buitelaar, E. 2011. Institutions in theories of land markets: illustrated by the Dutch market for agricultural land. *Urban Studies* 48(1), 161–176.

Sorensen, A.D. and Day, R.A. 1981. Libertarian planning. *Town Planning Review*, 52(4), 390–402.

Velten, A.A. van 2003. *Privaatrechtelijke aspecten van Ontroerend Goed*. Deventer: Kluwen.

Webster, C. and Lai, L.W.C. 2003. *Property Rights, Planning and Markets*. Cheltenham: Edward Elgar.

Williamson, O.E. 1995. Transaction costs economics and organisation theory, in O.E. Williamson (ed.), *Organisation Theory: From Chester Barnard to the Present and Beyond*. New York: Oxford University Press, 207–256.

Williamson, O.E. 1999. Public and private bureaucracies: a transaction cost economics perspective. *Journal of Law, Economics and Organization*, 15(1), 306–342.

Chapter 2
Exploring the Effects of Property Rights Using Game Simulation

Esther Geuting and Barrie Needham

How the land in a particular location is developed depends on the laws, including the property rights, applicable there. We call this the 'local regime of land laws', and it includes all the rights connected to the land, the distribution of those rights, and the way in which those rights may be enforced. This chapter investigates how possible changes to those laws might affect what is built. But how can the effects of a hypothetical 'local regime of land laws' be investigated?

This question is interesting and important for planners, because if the local regime of land laws can affect development, then changing that regime can be purposefully used as a way of achieving the goals of land-use planning. And in that way, planners can expand their 'tool chest'. In this chapter we investigate a particular way of changing that local regime, namely creating new property rights which can be traded. Using that as an instrument for land-use planning we call 'structuring the market', deliberately changing the way in which the market for property development works. This is not 'more market' or 'less market', but creating a different market, that is a market which works in a different way and produces different results.

The question remains: if this is an important question, it needs to be investigated, but how? In practice, laws cannot easily be changed, as an experiment to see what the effects would be. This contribution puts forward two other ways of carrying out this investigation, in the laboratory as it were, namely with 'thought experiments' and with game simulations. Each can be used separately, but in the research reported here they were used in combination. These research methods have been applied in practice, and the results are presented here, together with an evaluation of the experience with those two research methods.

The reasons for choosing these methods, and how they are applied, are described in more detail below. They produce a rich supply of empirical results. It must be said however that those results are insufficiently valid or reliable for taking the political decision: the market should actually be restructured in a particular way. Nevertheless, the methods are a fairly efficient way of coming to a judgment, in the early stages of considering a change in policy or legislation, about whether a particular change deserves to be investigated more deeply, by for example a pilot or an experimental law.

The Hypotheses

We expected that changing the local regime of land laws so as to structure the market differently would affect the market for property development as follows:

- it would change the way in which property rights were exercised;
- that would give rise to new markets and constellations of sub-markets in property rights;
- that would influence the way in which market actors (both public and private) structure the process of property development;
- which would influence the end result, the new development.

These hypotheses were tested by applying the two research methods to three cases, in each of which the local regime of land laws was changed. Each of those cases was derived from a topical discussion in the Netherlands about how to improve property development by changing the legislation in a particular way.

The Three Cases

Case 1: Creating a New Property Right, Namely the Right to Develop Land, a Right which is Separable from the Right to Own Land.

This has been suggested as a way of increasing the competition between property developers in the local housing market. At this moment, the competition can be very small, resulting – it is suggested – in commercial housing developers offering a poor value for money. (This is the case not only in the Netherlands, but in England too – see Barker 2004.) This problem was noted in a government report ('Nota grondbeleid', VROM 2001). Within a local housing market, only a few housing developers are active and developers try to acquire development sites at an early stage, as a way of excluding competitors.[1] If that competition is limited, in particular within one segment of the housing market (for example new owner-occupied housing in a particular price range), we can speak of a local monopoly or oligopoly. The case investigated here was whether the competition could be increased, and what the effects would be, if the law were to be changed so that a development right could be compulsorily split off from the right of ownership, whereby that right could be traded under certain conditions. A development right is a sort of ground lease, and what is investigated here is the possibility that the owner of the freehold right can be obliged to sell that ground lease. Such a

1 The problem is not that there is too little competition between housing developers nationally: that was investigated and found satisfactory by the Nederlandse Mededingingsauthoriteit – the Dutch watchdog against monopolies. But within a local market, the competition can be very limited.

recognition of a separate 'right to develop' could have practical effects similar to the nationalization of development rights carried out in England and Wales in 1947 (Booth 2002), although the legal basis would be totally different.

This case is an example of the market being changed directly, by introducing a new property right. In the following two cases, the planning law (public law) would be changed, which – it was expected – would then lead to changes in the way the market worked.

Case 2: The Redevelopment of Post-War Housing Estates by Introducing Transferable Development Rights.

Some housing estates built shortly after the Second World War are in need of redevelopment. But that is proving to be very difficult. It is not only that redevelopment itself is very expensive, but also that redevelopment usually results in less housing (certainly, in less housing floor space). This problem was signalled in two national reports: 'Tijd voor keuzes' (VROM-raad 2007) and 'Grond voor kwaliteit' (VROM-raad 2009). There is a change in the market structure which might help, namely creating saleable rights which could be exchanged between redevelopment areas and greenfield sites. This could be done by introducing contracts which could be entered into voluntarily between developers on the two sorts of development locations. Those contracts would become attractive by a change in planning law, whereby within the limits of the planning area (consisting of the redevelopment area and the greenfield sites) the saleable rights are the only available development rights in the market.

This is a form of market structuring which has already taken place in the United States and many other countries, in the form of transferable development rights (see Pruetz 2005 for a survey). The research that has been carried out in the Netherlands into this possibility so far has paid little attention to the economic effects and the consequences for the actions of market actors. And possible applications to property development have hardly been investigated, apart from the 'red for green rule' in Limburg (Provincie Limburg 2008).

Case 3: Land Readjustment as a Way of Reducing the Effects of Fragmentation of Land Ownership on Development Locations.

If land ownership is very fragmented, it can be very difficult if not impossible for private developers to carry out the development (Buitelaar 2008; VROM-raad 2009). In such situations in the Netherlands, it is the practice that municipalities themselves acquire all the land and other property, and then make the site suitable for building (changing the boundaries of the land parcels, putting in new infrastructure, etc.), and then sell the new building plots to private developers. This is, however, very expensive and involves the municipality in a huge financial risk. In this case, we investigate the possibilities for making such development sites profitable without municipalities first taking the land into ownership. This would require a form of urban land re-adjustment. The case investigated here is a change in planning law (public law) which would then stimulate the creation of readjustment rights which could be traded privately.

This has been possible in Germany for more than a hundred years, and there it is one of the most important instruments for urban development. It has been considered in the Netherlands since the 1940s. After the Second World War, that was for the rebuilding of housing and town centres destroyed by bombing (and it was used for the reconstruction of the centre of Rotterdam). In the 1970s and 1980s, land readjustment was to make urban renewal easier and quicker (De Haan 1985; Van der Putten et al. 2004; Needham 2007). In the government report 'Nota grondbeleid' (VROM 2001) the possibility was mentioned of creating temporary land banks in redevelopment areas, as a way of stimulating land readjustment and redevelopment, and more recently the ministry has been considering this.

In all three cases, the necessary law does not exist and is it not expected that it will be created, certainly not in the short term. Nevertheless, the problems are widely recognized, as are the possible legal changes which we investigate, and they are recognized not only by public bodies but also by the commercial parties active in property development. The cases are investigated in order to illustrate the possibilities of 'market structuring' as a policy instrument in land-use planning, and as a way of investigating those possibilities using thought experiments and game simulations.

Thought-experiments as a Research Method

The thought-experiments try to answer the question: what would be the probable effects of a certain law or a certain change in the law? That answer can, in principle, be tested in practice, but only if the law has, indeed, that content. The method followed here is to create an imaginary situation, very carefully specified, whereby the local regime of land law is changed in a particular way; to reason through the possible consequences (what-if scenarios); and to ask experts to comment on those scenario's. Those experts were lawyers, land developers and academics. When predicting the reactions to the new law, a distinction was made between the reactions of:

- the property developer;
- the builder;
- the investor in land and property;
- the housing association;
- the first owner of the development land, or the supplier if that were different;
- the municipality.

All those actors are involved when new housing is built.

This method was followed by Needham and Geuting (2006) in order to predict the effects of abolishing the automatic right of the owner of land with planning permission to develop his land, a possibility which was being investigated by the parliament at that moment. It is based on ideas from law and economics: using economic theories, the consequences are predicted and/or evaluated of a particular legal decision, a series of decisions, or a legal ruling (Teijl and Holzhauer 1995: 318). The method of thought-experiments falls within the 'positive theoretical approach' of law and economics. This aims for an objective explanation (better: an inter-subjective explanation, one that is shared by a large number of persons) of the law (Teijl and Holzhauer op. cit.: 317).

When applying thought-experiments for this research, use was made of economic theories such as developed in law and economics. Cooter and Ulen (2001) and Posner (1993) give more information about the practical application of economic theories to legal questions. Harvey (1987) did something like that many years ago, applied to questions of real estate. That application of theories was complemented with knowledge, gained from many years consultancy practice in the world of real estate, about how market actors behave, and with the reactions of consulted experts.

Game Simulation as a Research Method

It was decided to complement the method of thought-experiments with game simulations. The reason was to gain more and better insights into how market actors would react to the possible changes in the local regime of land laws. In particular, the thought-experiment carried out by Needham and Geuting (2006) resulted in the expectation that small differences in the way that the legal ruling is formulated can have big consequences for how market actors react: but that method could go no further. The strength of game simulation is that it can provide additional insights into the ways of thinking and acting of the public and private actors. There are other research methods which are sometimes applied to predict the consequences of such changes, such as interviews, brainstorms (the Delphi method, group decision rooms, etc.), but these are not seen as being very reliable. That was the reaction too of the participants in the game simulation, many of whom had experienced those other methods. The participants were asked to evaluate the game simulation afterwards, and they said that it was instructive and inspirational to see from close by how the other market actors (often competitors) made strategy and chose tactics. The participants said that they now understood better not only the possible effects of changing the legal rules, but also how the market for property development worked.

More generally, in comparison with other research methods, game simulations seem to approach better what the results of actually changing the law would be. Mastik et al. (1995: 119) say that game simulation is an appropriate research method for investigating possible legal changes in the following circumstances:

- a complex environment;
- external factors exert a big influence;
- a large degree of uncertainty;
- many actors and many reciprocal relationships;
- the actors do not all have the same influence;
- there is a variety of opinions about problems and solutions;
- the persons and organizations involved will anticipate the change in law and act accordingly.

These circumstances are common in situations where real estate laws and – in particular – land-use planning are involved.

With game simulation it is possible to take account of the practical experience of municipalities, developers and investors with developing new housing, and to apply that to how they might work with new legal rules. This could be an important contribution to academic research in planning.

The main aim of the research method was to register the reactions to new laws. This concerns not only whether the intended and predicted effects are realized, but also the unintended and unforeseen effects. Game simulation is particularly good in this (Mastik 1995: 15). One reason is that with it (unlike other methods like brainstorming and thought-experiments) the participants are directly confronted with the new legal rules. Moreover, the simulations require that participants interact with each other in their various roles. It is a sort of dynamic evaluation instrument. This is particularly important when predicting the effects of new planning instruments, for there are many different actors each with their own interests and aims, and their actions are highly dependent on the actions of others.

In this research, the thought-experiments were a useful preparation for constructing the game simulations, and the game simulations were used to test and to deepen the results from the thought-experiments. If the games gave results different from those of the thought-experiments, the results from the games were considered to be better. First, because many more persons were involved in the games than in the experiments, and second because the results came out of interactions and for that reason can be expected to be nearer to what would happen in practice.

Setting up the Game Simulation

The three possible changes in 'local regimes' were translated into three games, each to take place in the same fictional world. This was a development location called Iederpolder (everyone's polder) in the municipality of Honderduizendstad (the hundred thousand town).

It is important that the simulation represents, in a good but simplified way, the process of developing housing in a particular location. The changes in the law must be fully and clearly explained. And there must be a clear description of the context: the formal and informal institutional structure of the property development on that location, and the market circumstances under which the actors work. That latter includes the prices of rights in land and property as they now are (that is the context with which the actors are familiar), the scarcity of development sites, the relationship between new housing and the existing housing stock, and so on.

The most important steps in setting up a game simulation are, according to Duke (1980), Mastik (1995) and Peters (1994) the following.

Step 1: Making a System Analysis.

Aspects of the way property development works in practice must be 'frozen' (Mastik 1995) to determine the initial situation. The starting point was to select those aspects of the system in which we were most interested. These were:

- the market for land and the market for building development;
- the amount and the rate of transactions in the land market;
- contractual agreements;
- the price formation of property rights;
- the effects on the value for money of the end product;
- the effects on the variety of houses built.

Further, the actors were characterized very simply according to:

- the most important aims of the organization (for example commercial housing developer, housing association);
- the financial capacity of the organization (big or small);
- the type and extent of the property rights owned at the beginning of the game.

And the initial situation was further specified according to:

- the formal institutional structure of the housing market;
- the informal institutional structure of the housing market;
- the properties of the housing market (prices, composition etc.) and the contextual factors which influence this (such as the macro-economic situation);
- the situation in the fictional municipality, including the local housing market;
- the situation in the fictional location, including the division of land ownership.

Then these elements were related to each other. Examples are the responsibilities between the actors, the exchange of rights, flows of money and of information, causes and effects.

Step 2: Setting up the Environment for the Simulation.

The sorts of results which the game simulation must produce were specified, in relation to the aims of the research. Also, it was worked out what this meant for the simulation (the problem definition): what conditions must the game satisfy? An important choice was the level of abstraction: how abstract, concrete or realistic must the environment for the simulation be? This choice was made taking account of the wish to make use of the knowledge and skills which the actors (all from practice) had built up in their daily work. We chose a fairly simple environment and a fairly low level of abstraction.

Step 3: Designing the Details of the Game

The following components of the system were chosen:

- *The actors and what they must do*: They must develop 1000 dwellings on a particular location, under specific market circumstances, with specific market actors.
- *The events, what happens during the simulation*: In particular, during the simulation the legal rules would be changed in a particular way.
- *The roles*: These were: commercial developers, investors, housing associations, land owners, officers of the local council, together with the experience which the participants have and/or the function which they perform.
- *The cycle on the day of the game*: This was a continuous process, and every hour one (game) year passed.
- *The rules of the game, supplementing the above details.*

Step 4: Building the Game

In this stage, all the necessary material and documents were made, the roles were described, and a scenario for the day itself was written. For example, a booklet was made in which the Honderduizendstad was presented, with its mayor, statistics about the housing market, price trends, facilities, and images of the possible new development Iederpolder. The change in legal rules was described in a brochure, and was announced in an article in the *Poldercourant* (local newspaper) which the participants received in advance.

Playing the Game

During the /game simulations, the participants had to work with the new legal rules in a simulated environment, being in possession of all the relevant information, and using their own background knowledge. The questions that arose they discussed in a practice situation, while trying to develop an imaginary housing location

How the Game Simulation Worked

Each simulation lasted one day. It consisted of two short plenary sessions (in the beginning, in order to explain the scenario, and at the end, to evaluate the game), and a continuous group session in which the participants tried to develop 1000 dwellings on a particular location within a local legal regime which changed during the process. The participants experienced the strong and weak points in the legal change and got a feeling, from the interactions between the market actors, for what the legal changes could mean for them and the other market actors

Registering the Course of the Game

During the game, support was given by two 'leaders' and six note takers. All the results, including the discussions and the decisions, were recorded with sound recorders and by the note takers. In spite of the fact that this registration was done as factually as possible, it is possible that it was influenced by interpretations. This is unavoidable. The aim of the registration was to reconstruct the day as precisely as possible, using the notes of the 'secretaries' and the sound recordings. In order to be as objective as possible, on many occasions several note takers were present.

Selecting the Participants

There were 15 to 20 people present for each game, fulfilling the various roles. These participants were selected as being people with experience of that role, using what is called 'theoretical sampling'. (This is, according to Peters and Van de Westelaken (2005), an appropriate method for selecting respondents with a specific expertise.) The researcher sought, directly, people who could fulfil the specific roles, either because they did that work now or had done it for a long time previously. Especially attractive were those who had, in the course of their career, fulfilled more than one of those roles, for then they could anticipate better the reactions of the other participants.

The participants were not selected randomly, but on the basis of their expertise and skills. That requires from the researcher, 'that you look at the real practice as openly as possible, searching for established but also for contrasting cases' (Peters and Van de Westelaken 2005). In this way, the researcher selected first from people in her own network, secondly she asked those people about other possible participants in their network, participants with a certain experience and function. The aim was that the participants should, so far as possible, be able to represent market actors in property development. That meant, in practice, well-educated and professional men and women with at least several years of work experience in one or more of the roles. The participants had, in general, an important position within their own organization, but they did not sit in the management. They were not selected on the basis of any other characteristics.

There are, of course, others groups (roles) active in property development, besides those selected for these simulations. There are the financers and people who, within their own organization, can influence the go/no-go decision to purchase land or to develop a particular location. There are (other) advisors, there are lawyers, and there are end-users (the future occupants of the dwellings). None of these was included in the game simulation. The reason is the expectation (of the researcher but also of the other market actors) that these persons are not active in the core of the development process, nor do they exercise much influence on the process as a whole. This was confirmed in the evaluations carried out at the end of each simulation.

Testing the Hypotheses

Above we formulated four hypotheses about how structuring the market differently would affect the market for property development. Here we compare those expectations with the results from the thought-experiments and the game simulations.

The first hypothesis was that changing the market structure would change the way in which property rights were exercised during property development. This turned out to be correct, to a greater or lesser extent, in all three cases. However, what also appeared was that the existing informal practices remained in force and had a big influence on how the actors reacted to and used the new legal rules. These existing practices are the customary working procedures in the market for property development, and they have become established in the course of many years. Examples are the tendency of municipalities and developers to work together, also the cooperation between housing associations and between land owners with shared interests. Sometimes this produced results different from those predicted from theories in law and economics. This outcome underlines the added value of using not just thought-experiments but combining them with game simulations.

The second hypothesis was that the change in the way in which property rights were exercised would give rise to new markets and to constellations of sub-markets in property rights. This was only partly confirmed. For the resilience of customary practices influenced this too: the choice of partners appeared to be the same as that under the existing rules. This can be summarized as: big organizations prefer to work with other big organizations. Further, the actors did not want to perform other, new, tasks. Builders remained builders, and investors remained investors, in spite of the expectations that the new legal rules would make it profitable for them to change. Developers, municipalities and housing associations appeared to continue to work with the actors with whom they were already familiar. And other actors, who might have wanted to get into those sub-markets, found that very difficult.

This can be interpreted in the light of the statement by Coase (1960) that the initial assignment of rights has a large influence on the outcome. The division of property rights between the market actors at the beginning of the simulation affected how the actors reacted to the change in legal rules. In particular, those with the most rights initially have a big effect on the outcomes.

The third hypothesis was that changing the market structure would influence the way in which market actors (both public and private) structure the process of property development. This was confirmed in all three cases: but the effect on the development process was not always that which had been expected. In particular, it appeared that changing the local regimes might very well delay the process, at least in the short term, because of the necessity to get used to the new rules.

The fourth and final hypothesis was that the above changes would influence the end result; that is, the new development. That too was confirmed, but here too the effects were not always the same as had been predicted.

Conclusions about the Two Research Methods

The experience with using those two research methods – thought-experiments and game simulation – enables us to draw another type of conclusion, namely about the usefulness of those two methods in research for planning.

The Strong Points with those Two Research Methods

The experience with applying these two research methods is positive and leads us to conclude that they could with benefit be used more often in planning research. Game simulations are one of the few ways which are appropriate for ex-ante evaluation of possible changes in policy and laws and of changes in planning instruments. Also, they could be used for ex-durante evaluation (continuous monitoring) of changes which had been introduced.

People already working in property development – project managers and policy assistants from municipal development agencies, commercial project developers, builders, housing associations, investors – could also learn from game simulations. For that could increase their understanding of how the development process works. The same applies to those training to work in those professions, for the methods show them the various roles in the development process and how they can be fulfilled.

More particularly, the thought-experiment, and even more the game simulation, makes clear the differences between the actors in the development process. Sometimes, people talk as if developers, investors and builders were indistinguishable. But the simulation shows that they have different interests, that they act differently, and that they experience different consequences of changing the rules.

Secondly, it became clear that the expectations held by the participants at the beginning of the game about the effects of the new legal rules differed from the experience they gained during the game. For in the game, they experienced directly (albeit in a simulated environment) what those new rules could mean in practice. A good example is how the representatives of municipalities viewed the splitting of development rights from ownership rights. Initially, those representatives were enthusiastic supporters of the change, based on the policy discussions about this possible change and on their practical experience with how developers on greenfield sites try to exclude competition. In the course of the simulations, however, they became sceptical about whether more suppliers on the local housing market would effectively increase the competition between them.

Thirdly, the game simulations showed how important the interactions between the actors are, and what a great influence this can have on what is actually built. In all three games it was clear how the market actors influenced each other, including the choice of partners, and how this influenced the price of property rights, the choice of development strategy, and the collaboration.

Further, it appeared that participants in a game simulation acted as they would in their daily professional work. Wherever possible, the participants were placed in roles where they could operate in a context with which they were familiar in their normal work. This made is easier to see how the representative of a particular role might react to the new legal rules. This was helped by the realism of the game environment, but more by the setting on the day itself, whereby market actors continually made contacts with each other so as to conclude transactions and agreements.

Another strong point of game simulation is that you can observe how people act, whereas in interviews and brainstorms, it is more a question of reasoning and arguing about the possible effects of changing a local regime. In the game simulation, the changes were made very tangible, whereby the participants had to make use of their own expertise.

Finally, the game simulation showed that both public and private actors have much freedom in how they react to new rules. These actors bring matters up which are important to them, matters which the researcher had not mentioned. This shows the importance which the participants themselves attach to the various issues. For example, it appeared that the effect of the changes in legal rules on the 'value for money' given by the new dwellings attracted less attention than the way the price of property rights was set and the speed of development.

The Weak Points with the Two Research Methods

The weakest point with game simulations is the selection of the participants. A large number of people have to be found who are willing to devote a whole day to the research. This requires from the researcher a very good network of professionals. In addition: in order to select suitable people you need information about their backgrounds. Moreover, the selection that is made can have consequences for the results: and the selection is not made at random. Even one person can affect the results. And the quality of the participants affects the quality of the outcomes. This methodological limitation means that the results of the simulations must be accepted with the necessary reservations, including reservations about the validity and generalizability of the findings.

Another disadvantage of game simulations is the reverse of one of the advantages: the researcher exercises little direction during the simulation itself. The registration provides much information which is not relevant to the research. And there is no possibility, as there is with an interview or brainstorming, to ask additional questions about a particular topic, so as to get a better understanding of it.

There are practical disadvantages too to game simulations. It takes much more time than interviews or brainstorms to prepare and develop the 'environment' and to create a good setting. For that reason, game simulations are expensive. Nevertheless, if the alternative were to introduce a new experimental law without the necessary preparation, that could be much more expensive!

Carrying out a game simulation is risky. It must take place in one day, and participants might withdraw at the last moment (as they did in all three games). Game simulations are exciting and stressful for the researcher!

Limits to the Validity and Generalizability of the Results from Game Simulations

Neither the participants nor the researcher nor the assistants can have a complete overview of what is happening during the simulation. That means that those who register the results do so from their perception and interpretation of the course of events: they cannot fail to reconstruct those events, so they create as it were a new reality (Mastik et al. 1995: 73). Using a large number of observers can help to get a record which is as objective as possible.

In addition, the sound recordings were a good way of finding what was actually said during the simulation, and in that way checking the interpretations. And two times during the simulation, the participants were asked to fill in a questionnaire about the expectations which they had about their own actions, the actions of others, and the effects of those. In those ways, it was attempted to strengthen the internal validity of the registration. It is good to note that the participants, when asked to evaluate the environment within which they had to play the game, said that it was a realistic simplification of the reality they knew.

In this research, each case (a particular change in the local legal regime) was investigated with one game simulation. As a result, the generalizability of the results is limited. It would have been better to carry out more simulations, under different circumstances and with different actors: but for practical reasons, that was impossible. Nevertheless, that limitation is acceptable for research which aims to explore a new situation. If the aim had been to 'test' different scenarios in preparation for a change in legislation, the limitation would have been serious. But for this research, the three cases were chosen to illustrate the possibilities of a type of instrument (market structuring), not to test the applicability of a particular legal regime.

In spite of that, the results booked here can contribute to discussions about market structuring in land-use planning. So it is recommended that the game simulations be repeated in order to see if the results are robust and do not change under the various circumstances that are found in property development. In that way, it can be seen to what extent the results can be generalized.

It is to be Expected that the Following Circumstances would Influence the Results

The market situation The game simulations were carried out in the summer of 2007, when the property market was such that all participants had rosy expectations about the market for real estate and for new housing. A few years later, the expectations of both private and public parties have altered drastically, as a result of the credit crisis and the economic recession. In a follow-up, the environment for the simulation could be changed to include falling prices and changes in the demand for housing.

Relationships within the market The game simulations started from a certain distribution of legal rights between the participants. Would the effects of changing the legal rules be different under a different distribution of rights? Considering the transaction costs, one would expect that the initial assignment of rights, including the volume of rights possessed by actors, would affect the results of working within a given law regime. But how great would those effects be? And would this influence the effectiveness with which a purposeful change to the local legal rules achieved its aims?

The structure of the market Based on the work of Needham and Geuting (2006) it is to be expected that the results of the game simulations are significantly influenced by the details of the specific rules of the game. Small changes in the property rights regime might lead to big differences in the outcome of the game simulation.

Time effects It is probable that the reactions of stakeholders would change over time. In the short run, it is to be expected that the reactions are strongly influenced by existing working practices. In the long run, practitioners have more opportunity to think about new possibilities offered by the new property rights regime, and so to adapt their practices accordingly.

Conclusions about the Practical Relevance of these Research Methods

The experience with applying the combination of thought experiments and game simulation was encouraging. It indicates that the consequences of a possible change in legislation for land use or for property rights can be forecast fairly reliably. If the methods were applied, they could help to prevent failed legislation: that is legislation – like the development charge introduced in England and Wales in 1947 – which does not produce the desired effects. And more widely, the methods could be applied for forecasting the effects of new planning measures taken within the existing legislation, such as innovative approaches to urban regeneration or to financing large developments

References

Barker, J. 2004. *Review of housing supply: final report, recommendations.* [Online]. Available at: www.barkerreview.org.uk [accessed 22 September 2011].

Booth, P. 2002. From property rights to public control; the quest for public interest in the control of urban development. *Town Planning Review*, 73, 153–169.

Buitelaar, E., Segeren, A. and Kronberger, P. 2008. *Stedelijke transformatie en grondeigendom.* Den Haag / Rotterdam: RPB / Nai uitgevers.

Coase, R. 1960. The problem of social costs. *Journal of Law and Economics*, 3, 1–44.

Cooter, R and T. Ulen 2003. *Law and Economics*. Boston: Pearson Addison Wesley.

Duke, R. 1980. A paradigm for game design, simulation & games. *An International Journal of Theory, Design and Research*, 11(3), 364–377.

Harvey, J. 1991. *Urban Land Economics, the Economics of Real Property.* Basingstoke and London: Macmillan Education Ltd.

Haan, P. de 1985. *Voorontwerp tot wijziging van de Wet op de stads- en dorpsvernieuwing, hoofdstuk IVa Stedelijke herverkaveling.* Delft/Amsterdam.

Mastik, H., Scalzo, R.G., Termeer, C.J.A.M. and in't Veld, R.J. 1995. *Simulatie van Wetgeving, een verkenning van gebruiksmogelijkheden van spelsimulatie als ex- ante evaluatie van wetgeving*, Achtergrondstudies Algemeen Wetgevingsbeleid. Rotterdam.

Needham, B. 2007. The search for greater efficiency: land readjustment in the Netherlands, in Hong Yu-Hung and B. Needham (eds), *Analyzing Land Readjustment: Economics, Law and Collective Action.* Cambridge: Lincoln Institute of Land Policy, 115–134.

Needham, B. and Geuting, E. 2006. *Afschaffing van het zelfrealisatiebeginsel, economische effecten, deelopdracht in opdracht van het ministerie van VROM.* Nijmegen: Radboud Universiteit Nijmegen, Faculteit der managementwetenschappen, onderzoeksgroep Governance and Places.

Peters, V. 1994. *Spelsimulatie & gaming, systeemanalyse: ontwerpen van onderzoek*, Nijmegen: Katholieke Universiteit Nijmegen.

Peters, V. and van de Westelaken, M. 2005. *Spelsimulatie & gaming, checklist voor het opstellen van de specificaties van ontwerp*. Nijmegen: Samenspraak Advies.

Posner, R.A. 1993. The new institutional economics meets law and economics. *Journal of Institutional and Theoretical Economics*, 149(1), 73–87.

Provincie Limburg 2008. *Monitorrapportage ruimtelijke kwaliteitsinstrumenten.* Maastricht: Provincie Limburg.

Pruetz, R. 2005. *Beyond Takings and Givings: Saving Natural Areas, Farmland and Historic Landmarks with Transfer of Development Rights and Density Transfer Charges*. Marina Del Rey, CA: Arje Press.

Putten, E. van der, de Lint, R. and de Wolff, H.W. 2004. Stedelijke herverkaveling: Motor voor vernieuwing van particulier bezit. *Tijdschrift voor de Volkshuisvesting*, 10(5), 27–31.

Teijl, R. and Holzhauer, R.W. 1995. *Wisselende perspectieven in de rechtseconomie*. Deventer: Gouda Quint.

VROM, Ministerie van 2001. *Nota Grondbeleid*. Den Haag: VROM.

VROM-raad 2007. *Tijd voor keuzes, perspectief op een woningmarkt in balans, VROM-raad*. Den Haag: advies 064.

VROM-raad 2009. *Grond voor beleid: Voorstellen voor verbetering van overheidsregie op (binnen)stedelijke ontwikkeling*. VROM-raad. Den Haag: advies 070.

PART II
Coping with the Wider Institutional Framework

Chapter 3

Planning Law Reform and Change in Post-apartheid South Africa

Mark Oranje and Stephen Berrisford

Over the last seventeen years, much has been said and written about the need to radically transform the deeply unequal and divided South African society and the country's settlement pattern that not only reflects, but also entrenches and enforces it. While significant changes have been made in the country's institutional superstructure, the material conditions on the ground have proven to be far harder to change. In addition to this, components of the superstructure that could potentially assist in ensuring the latter, such as planning law, have proven resistant to change. Despite this, a strong view remains that a new planning law can, and will make a major impact in the pursuit of societal transformation.

These introductory comments raise many questions, such as: Does this positive view of the transformative power of planning law have merit? What is the view based on? Is it located in a belief in the power of planning law to create and sustain economic value through the allocation and regulation property rights? Following on from this, if it was perceived in this way, why did the State not do more to see new planning legislation passed? What prevented it from seeing through the desired changes, and why is planning law reform so difficult to push through?

In this chapter, we seek to engage these questions through an exploration of the post-Apartheid endeavours at producing a new planning law-framework for the country. In this process, we explore the challenges the process encountered and the results/impacts of the changes that were made.[1] While not comprehensive, it does provide a base to enable at least a tentative engagement with the issues as raised. In addition, it provides a case-study for those interested in transforming planning law for the purpose of using it to realize other objectives, for example environmental sustainability.

The chapter has four parts. The first provides a brief overview of planning and planning law in South Africa, pre-1994. The second tells the story of the route travelled with regard to planning law over the last seventeen years. The third discusses elements of the post-1994 route, seeks to make sense of what happened and engages the questions as raised above. The last provides a series of conclusions.

---

1 For a more detailed analysis of the process of post-apartheid planning law reform see Berrisford (2011).

Planning and Planning Law circa 1994

When South Africa's first democratically elected government took office in May 1994, it found a planning system and planning law framework that had for more than 150 years been intertwined in the economic objectives and ideological ambitions of successive colonial and Apartheid regimes.[2] As this framework is the one that the new government sought to change, a few notes on its history, characteristics and functioning are of value.

Planning legislation (in name) was first introduced in South Africa in the early 1930s. This followed a long struggle by a small group of visionary urbanists, primarily architects, to overcome strong property sector interests and drawn-out intergovernmental tussles as to where such legislation should be located. The system made provision for orderly procedures to be followed when (1) establishing new settlements, and (2) allocating and changing the development rights on properties. This was a far cry from the modernist, utopian ideals of many of its early proponents, which included comprehensive futurist plans and targeted intervention for the public good (Oranje 1998).

Given the country's status as a British colony until 1910, and a Dominion between 1910 and 1961, it was strongly influenced by similar legislation passed in Britain. In essence, conditions in that country were the main point of reference.[3] At the same time, planning law was from the outset conceived and promulgated for exclusive use in the 'European/White areas' in towns and cities.[4] The comprehensive and enforced planning system established in this way, enabled (1) the orderly development of towns and cities in 'White South Africa', and (2) the creation and protection of property wealth in these areas. In 'areas set aside for African occupation', rudimentary land-use planning/parcelling was done; nothing that even remotely resembled the detailed regulatory systems put in place in White areas, even after more sophisticated planning systems were instituted in 'peripheral urban areas set aside for Black people' in the mildly reformist 1980s and early 1990s.

The decades of macro-segregation, separation and the 'cutting out' of 'Bantustans/homelands' from 'white South Africa' from the early 1970s onwards, resulted in the creation of a complicated legal framework. The four provinces created in 1910, and the ten Bantustans over the next eighty years, each had in

---

2 For a historical overview of the system as it developed pre-1994 see *inter alia* Mabin and Smit (1992); Mabin (1991); Muller (1983); Oranje (1998).

3 This meant that the legislation was not geared towards conditions in South African, given that it was conceived with social and economic and conditions in urban, industrial Britain as setting.

4 Areas in towns and cities reserved for settlement by Africans were from the early 1920s onwards managed by the Native Urban Areas Act, 1923, and amendments to this Act (Oranje 1998). For more on the origins of planning legislation and its links with racial segregation see *inter alia* Parnell (1993) and Oranje (1998).

place a different planning law system. When nine new provinces were created in 1994, provincial and local authorities were tasked with administering in some cases up to seventeen pieces of planning legislation, each with different application procedures, terminologies, definitions and submission and approval mechanisms (Oranje 2002). As part of the agreements reached in the run-up to the 1994-elections by the then governing National Party and the African National Congress (ANC), any law in existence before the political transition remained in force afterwards in relation to the geographical area to which it had applied beforehand. In doing so, a situation of legally sanctioned, unequal treatment was established, as the legislation for the former 'White areas' provided far more public protection and community participation than the legislation in the former 'African areas' did.

In most municipalities, urban planning was located as a sub-section in City Engineers' Departments (Oranje 1998). Situated in these technical departments, planning was essentially called upon to do two things: (1) development control through the regulation of land-use and development rights, and (2) so-called 'forward/master planning'. The latter was instituted in the 1970s in the larger towns and cities, and typically entailed the production of static blueprint-like master and structure plans. These plans had limited powers of enforcement, were generally not considered in municipal budgeting and were primarily used when engaging prospective developers on development applications. These plans were not treated with much respect by developers or other professionals, leading in many municipalities to the gradual marginalization of plans and their drafters and custodians, that is, the planners.

The introduction of planning legislation in the country was accompanied by the emergence and development of a distinct planning profession. Whereas 'planning work' (for example settlement layouts and applications for land development rights) was in the first half of the twentieth century largely done by architects and land surveyors, a new planning profession emerged in the 1950s. New on the scene, much smaller than the older established professions (notably the architects and the engineers), and eager to be taken seriously, the profession adopted a low-key, subservient role. As such, it was an exercise in 'pleasing' – both the State and the private, property development sector. This resulted in planners expediently focusing on the technical detail of their work and steering clear of controversy. This meant that they did not speak out *en masse* against the implications of Apartheid for the lives and life chances of Black South Africans in peripheral, rudimentary and under-serviced settlements throughout the country. In a far cry from the reformist roots of planning in notably Great Britain, planners acted as cold, task-oriented bureaucrats. They were cogs in a machine that was very good at keeping things as they were, not at changing them. This bureaucratic, apolitical stance met with growing opposition from progressive individuals and NGOs especially from the 1980s onwards, which by the early 1990s had resulted in a deep fracture in the profession. This, together with the scorn heaped on it for its unquestioning involvement in the Apartheid system,[5] saw the planning profession weak and divided at the transfer of power in April 1994.

Planning Law Post-1994

While there were many uncertainties in the early days of the new dispensation, there was no lack of certainty that the pre-1994, differentiated planning system had to be replaced with 'one equal planning system for all'. On the specifics of this new system there was less clarity. Should there for instance be one national planning Act, or one national Act operating in tandem with nine provincial Acts, or only nine provincial Acts? It was hoped that this would be resolved through the Constitution-writing process underway at the time, but this was not to be.

The final 1996-Constitution addressed the issue of which 'sphere' of government – national or provincial – would exercise particular legislative competences. However 'town/urban planning' was not included in the crucial lists of law-making functional areas.[6] The lists *did* include other areas for law-making such as 'provincial planning', 'regional planning and development', 'municipal planning' and 'urban and rural development', but did not clarify how these related to the existing structure and activity of 'town and regional planning'. This resulted in legal uncertainly as to whether national or provincial government was charged with the responsibility of planning law reform.

Amidst these uncertainties, the national government passed new legislation in October 1995, the Development Facilitation Act ('the DFA').[7] The DFA signalled a major departure from the pre-1994 planning legislation in a number of ways: Firstly, the Act was grounded in and included a set of normative principles with which all development and planning activities had to comply. These principles centred on the need to remove the spatial inequities created by apartheid and as such signalled a transformative role for planning. Secondly, the Act instituted strategic planning in the local government sphere by requiring all municipalities to prepare Land Development Objectives (LDOs) on a five-year base, and to conduct annual reviews of these 'plans'. The LDOs effectively became the first post-apartheid municipal planning instrument. Thirdly, the DFA provided for provincial Development Tribunals as an alternative decision-making body on land use and land development applications. The choice whether to submit an application to a Development Tribunal or the municipality lay with the applicant. These Tribunals had extraordinarily wide powers, including the right to suspend the operation of

5 The then Chair of the South African Council for Town and Regional Planners, the late Prof John Muller, in 1997, made a submission to the Truth and Reconciliation Commission, in which he formally apologised on behalf of the profession for its unquestioning complicity in the Apartheid system (see Muller 2003).

6 Schedules 4 and 5 of the 1996 Constitution.

7 The Development Facilitation Act, No 67 of 1995.

a number of planning, environmental and land administration laws that would otherwise have had a dilatory effect on the application. The Tribunals were also empowered to summon witnesses to appear before them, including government officials. Almost the only constraint on the Tribunals' discretion was that their decisions had to be consistent with the applicable LDOs. Finally, in step with its interim status, the Act provided for the establishment of a national Development and Planning Commission (DPC)[8] on which planning and development experts and activists would serve. This body was charged with researching alternatives to the inherited legal and policy frameworks for urban (and rural) development and planning, and advising the new government on reform in these areas.

The implementation of the DFA had barely begun, when the Local Government Transition Act Second Amendment Act, No 97 of 1996[9] was passed. This law placed a legal requirement on all local authorities to prepare Integrated Development Plans (IDPs). The linkage of these Plans to the LDOs was clear, with this Act defining an IDP as

> a plan … which has been compiled having regard to the general principles contained in Chapter 1 of the Development Facilitation Act, 1995 (No 67 of 1995), and where applicable, having regard to the subject matter of a land development objective contained in Chapter 4 of that Act. (Republic of South Africa, 1996: 4)

This period (the late 1990s) also saw a flurry of *provincial law-making*. In the case of Kwazulu-Natal, this went smoothly, resulting in the Kwazulu-Natal Planning and Development Act, No 5 of 1998, in June 1998. In the Western Cape, it was just the opposite with the initial conflict around the DFA escalating to what a regional newspaper referred to as a '*constitutional battle*' (Oranje 2002). After much heated debate around the Bill, the Western Cape Planning and Development Act, No 7 of 1999, was eventually promulgated in April 1999. In the primarily rural Northern Cape, the provincial legislature not only succeeded in enacting its Planning and Development Act in 1999 but, uniquely at the time, in bringing the new law into operation. In the Gauteng Province, a bill was completed in 1999, but failed to make the final transition into legislation until 2003. A Development

8 Before its disbanding in 2000, the DPC produced a draft Green Paper on Planning and Development, which the Minister subsequently published for comment. The White Paper on Spatial Planning and Land Use Management that emerged in 2001 out of this process was subsequently approved by the national Cabinet. In contrast to the Green Paper, this White Paper called for a much stronger role for national government legislation and the rationalization of the provincial laws into one uniform set of national rules and policies. In accordance with its preference for a single national Act, the Department published the first of many versions of a national Land Use Management Bill (LUMB), in July 2001.

9 This was an amendment to the Local Government Transition Act, 1993, which sought to provide interim measures for local government during the transformation process.

Planning Bill was drafted in the North West Province, but failed to get through the legislature. In the case of the Gauteng Province, the delay in getting the Act passed was said to be primarily a result of this province waiting for national government to finalize the Municipal Systems Bill, a new piece of legislation containing specific sections dealing with municipal planning (Oranje 2002).

At the same time, a range of national government departments embarked on a process of drafting new sector legislation, with the national Departments of Water Affairs and Forestry, Transport and Housing all promulgating new laws placing additional planning requirements on municipalities.[10] In addition to this, driven both by a series of local factors, and international wisdom at the time favouring decentralized, local planning,[11] the Ministry of Constitutional Development[12] produced a White Paper on Local Government in 1998. This White Paper articulated strong developmental and planning roles for municipalities. Subsequent to this, the Department embarked on the preparation of a suite of enabling and empowering local government legislation to give expression to that policy framework.[13]

As South Africa entered the first decade of the twenty-first century, the country had (1) a White Paper on the rationalization of planning laws, and (2) a draft Land Use Management Bill, in the public domain. At this point it seemed likely that the country would have a new land-use management system in place by the time of the celebration of the ANC's first decade in power in 2004. This, however, was not to be.

The first setback came with the new Land Use Management Bill itself, which was widely shunned. The reasons included: (1) its simplistic approach *vis-à-vis* the sophistication of the DFA and some of the post-1994 provincial bills and Acts; (2) the powers it sought to concentrate in the hands of the national Minister of Land Affairs responsible for the Act; and (3) the absence of an engagement with environmental aspects. At the same time, the Department of Land Affairs, the custodian of planning law and legislation regulating the planning profession, was going through a challenging period. To some extent, this was attributable to the department having lost many of its older White technocrats and the former activists

10 For example, in terms of the Water Services Act, No 108 of 1997, local authorities are required to produce Water Services Development Plans. Integrated Transport Plans are required in terms of the National Land Transport Bill, 1999 and certain requirements as to plans for housing provision are required in terms of the Housing Act, No 107 of 1997.

11 Notably (1) the new impetus given to decentralized, municipal planning with the 1996-Consitution providing for local government, alongside national and provincial government as a separate and distinct sphere of government, and (2) the closure of the national RDP Office in the same year, was followed by the handing of the mantle of pursuing the ideals of the RDP to the local sphere of government.

12 This ministry was subsequently renamed the Ministry for Provincial & Local Government, then the Ministry for Cooperative Governance & Traditional Affairs and now the Department of Cooperative Governance.

13 These Acts were the Municipal Demarcation Act, 1998, the Municipal Structures Act, 1998, and the Municipal Systems Act, No 33 of 2000.

that had joined its ranks in the first years after the democratic transition. Land reform, one of its key mandates, was also not proceeding as had been anticipated, which diminished its stature in the eyes of activists, researchers, the media and the public, as well as other national departments.

After roughly a decade in power (circa 2003/3) the focus in government under President Mbeki was shifting. In his earlier capacity as President Mandela's Deputy, Mbeki had been tasked with achieving greater intergovernmental cooperation. A key element of this entailed ensuring greater alignment in government expenditure to prevent State funding from enforcing the unequal Apartheid spatial model. This concern he brought with him to the Presidency, seeking ways in which a well-functioning system of intergovernmental planning, budgeting and implementation could be put in place. Financially too, the challenge of balancing government finances, something that was not treated with great urgency in the first five years of government, was making cooperative governance imperative. He essentially thus brought a different perspective to the business of managing the State, more pragmatic and more concerned with keeping the economy going.[14]

Provincially, dissatisfaction with the now long delayed new planning law was growing. Whereas many provincial governments had either delayed or suspended their provincial law writing processes, in anticipation of new national planning legislation, a number of them decided to restart their processes. This resulted in the passing of a Planning and Development Act in Gauteng (2003),[15] and the Western Cape and KwaZulu-Natal reviving their law-writing endeavours. Even in municipalities frustration was growing. A number of the larger municipalities, notably Ethekwini (Durban) Metro in KwaZulu-Natal and Mbombela (Nelspruit) in Mpumalanga, started exploring the option of using municipal by-laws to prepare their own land-use management systems, independently of provincial planning laws.

In practice, the lack of movement towards a new planning system had meant that a *de facto* dual planning system had been created, with both the old provincial planning ordinances and the Development Facilitation Act (DFA) available for use in most provinces. This meant that developers could choose to submit an application either to the municipality in terms of an old provincial ordinance, or to the province's Development Tribunal, in terms of the DFA. A key challenge with regards to the latter was that the DFA, enacted at a time when local government was mistrusted, gave the provincial Development Tribunal vast powers, even to over-rule municipal IDPs. This meant that developers could decide which legal route to pursue based on a variety of reasons, not least of which was whether or not the strategic plans of the municipality were supportive of their developmental intentions. On the other hand, the DFA provided a crucial support-function to

14 Many have labelled this 'a neoliberal turn' (see Oranje 2011).

15 Notwithstanding this however the 2003 Gauteng Act was never implemented. Within two years the province had commissioned consultants to draft a new bill. To date, mid-2011, no such bill has reached the legislature.

less-capacitated municipalities. This they did by diverting land development applications to the Development Tribunals who were significantly better able to manage and handle applications than the municipalities. While the DFA-option proved to be very popular in such cases, it was resented by the better capacitated municipalities as a direct threat to their constitutional power to plan and guide strategic land development, infrastructure and spatial planning. This frustration eventually led to the City of Johannesburg taking the matter to Court in 2007.

Amidst all of this legal wrangling, the Planning profession, after the fractures of the early 1990s, began to regain its footing (circa 2004/5). One of the contributing factors was the passing of the South African Planners Act, 2002 which *inter alia* prescribed the establishment of a new regulatory body, the South African Council for Planners. Despite this Council gradually gaining some trust amongst planners, and the South African Planning Institute organizing social events and hosting international conferences, the profession remained far short of the established and respected position which the Engineering and Architecture fraternities enjoy.

Throughout this period, the then Department of Land Affairs (currently the Department for Rural Development and Land Reform) continued to work on draft national planning legislation. This resulted in a bill being presented to Parliament in mid-2008, another Land Use Management Bill. The bill was widely rejected and sent back to the Department by the Parliamentary Speaker for further work. Simultaneously the City of Johannesburg's challenge to the constitutionality of the DFA was working its way up through the courts, finally reaching the Constitutional Court in 2009. In the June 2010 judgment, the Court confirmed that the Chapters of the DFA providing for Development Tribunals were indeed unconstitutional. It held that, because the Constitution guarantees local governments the power to exercise executive functions in relation to land use and land development, it was improper for a provincial Development Tribunal to do so instead.[16] In its ruling the Court found in favour of the City, confirming that both National and Provincial government can prepare and pass planning law, (1) subject to special rules governing a possible conflict between the two, and (2) provided that the new law concentrates decision-making powers at the local level. In addition, the Court stressed the need for far-reaching law reform and gave the national government two years to do this, setting a deadline of mid-2012.

With a judicial deadline looming, the Department of Rural Development and Land Reform published a new draft bill, renamed the Spatial Planning and Land Use Management Bill (SPLUMB) in May 2011 for public comment. The Court's two-year period for fundamental law reform, half of which had expired by the time that the comments on the draft SPLUMB were received, have put the Department under immense pressure to finalize the legislation for resubmission to parliament. As for the provinces, only KwaZulu-Natal and the Northern Cape have new, post-1994 Planning and Development Acts in place. KwaZulu-Natal's Act has been

16 The case can be found on www.constitutionalcourt.org.za. The citation for the case is *City of Johannesburg v Gauteng Development Tribunal and others* 2010 ZACC 11.

in operation since 2010, while the Northern Cape's law was enacted in 1998. As for the rest of the country, it remains a case of using the pre-1994 legislation, and waiting for the SPLUMB to work its way through Parliament.

Discussion

In the introduction to this chapter a number of questions about planning law and social change in South Africa were raised. These are returned to now.

The Power of Planning Law…

The initial overly-optimistic thinking amongst a dedicated group of development activists (circa 1993–1995) was that planning law could and would be restructured, and that this rewritten law would then lead to rapid and radical changes in society, economy and space. Planning law reform, however, very quickly became (1) entangled in the minefield of government institutions and inter-governmental dynamics; and (2) stalled by the realities of diverse and often-conflicting multi-actor motives and practices.

In the case of government institutions and intergovernmental dynamics, institutional turf contests between different spheres of government and even between departments and units in the same departments in the same spheres of government, was the first unforeseen and ugly hurdle. This was worsened by the inherent difficulty associated with seeing through a very sophisticated intergovernmental constitutional arrangement in a country undergoing rapid change in every area and sphere of its being. The lack of progress with the planning law reform was not only frustrating, but also resulted in this process being overtaken by other processes (for example new laws and policies in other sectors), leading to the continuous narrowing and curtailing of the manoeuvring space for a new planning law.

In the case of the agents, naivety and lack of experience on the side of former activists, coupled with a newness of officials to the task and gaps in their knowledge and skills-sets, led to numerous process-delays. This was worsened by inertia, complacency and an underestimation of the magnitude of the task at hand amongst both officials and politicians. Also, the drivers of the law reform processes were under-prepared for the resolve of the powerful vested interests in the property-owning and development sector to protect their interests. At the same time, strongly influenced by the introduction of the precepts of the New Public Administration in the latter half of the 1990s, notably performance contracts and rewards, risk-averseness amongst officials grew. Fearing personal failure became, in many cases, a stronger driver than wishing and working for the collective rewards that the future could bring. In addition to this, and set against the backdrop of deep poverty, the State became increasingly viewed not as a tool for societal transformation, but rather as a ladder for personal advancement and enrichment, securing a family income, and elevating an extended family from poverty. While clearly out of step with the more communal and collective tenets of the governing party, these self-centred pursuits were not countered by a more communally-focused national vision. The country was certain what it wanted to move away from – the Apartheid past – but it did not clearly articulate where it wanted to go to and how it could get there.

In addition, planning law reform was not assisted by the new State's *new* normative position. For a variety of reasons, the post-apartheid State had, over a period of less than four years after 1994, moved away from its initial discourse of reconstruction and development, to one of (1) 'not rocking the monetary policy-boat', (2) balancing the books, (3) keeping 'the international community'[17] content, and (4) steering clear of what were perceived to be 'risky endeavours'. The young, idealist and often naïve first post-Apartheid government was replaced by an expedient State increasingly adept at treading water at just the right pace and time. It could be argued that a focused, spirited planning movement could potentially have changed all of this, and have stirred the leaders and the discourse back into transformation-gear, but the low status and low profile of the planning profession, and the wish to survive the transition, dictated against this. The handful of activists in the broader development and planning fraternity, without the support of a wider grouping of planners, limited their inputs and attacks on policy positions to academic papers, and generally then, from the safety of the classroom, the liberal press and the conference venue.

On a more pensive and cautionary note, while it is true that (1) there were many challenges to counter in preparation and implementation of new legislation, and (2) this may be one of the reasons for planning law never getting the chance to play its (wished-for) transformative role, it needs to be questioned whether, even under ideal conditions, it would be able to do this. Simply stated: Would even 'the perfect planning law' lead to an improvement in the lives and life chances of the deprived African majority? Does planning law not tend to favour those that already have land and assets to offer as security? And, will the introduction of a post-apartheid planning Act provide the poor with properties on which to apply for settlement establishment, or changes in development rights? Continuing along this line, is a well-planned and executed 'land and other assets-expropriation and redistribution programme' not better suited for this task? Furthermore, and returning to the introductory chapter by Hartmann and Needham, could a new planning law, framed and 'managed' in a government system that had arguably lost sight of meaningful and sustainable progressive objectives, create and impart rights to 'the masses'? Would the system protect these frail, new rights and ensure

17 This 'grouping' of nameless, powerful outsiders and institutions who were (supposedly) closely monitoring events in the country and were ready to withdraw or withhold investments and funding from the country, has been used, whether valid or not for the last seventeen years, to see through often very contested decisions and policy directions.

that they were upheld both by communities and the State? Does neo-liberal elite formation and institutional lock-in, as the South African case has to date so clearly shown, not dictate against this?

There have been many notable and progressive changes in South Africa, such as improved access for previously disadvantaged groups to education and job opportunities, better housing and provision of municipal services, health care and social grants. These have however generally not been brought about by the use of planning law, or even by long-term municipal, provincial or strategic development plans. In most cases, they have been the result of direct material/tangible transfers, the removal of restrictions from the past, targeted programmes and funding schemes, and the introduction of fair labour practices and preferential procurement practices. If the question were not being asked from a planning law-perspective, as is being done in this chapter, planning law would in all likelihood not have been mentioned as a factor in a discussion on the drivers behind the positive developments highlighted above. Stated somewhat differently, what kind of rights, change and wealth *could* planning law have brought to the poor? What would the value and longevity of such rights be in a broader 'institutional and societal container/system' that had not been adequately prepared and transformed? At the same time, and with a broader agenda in mind, what comes first: transformation of the State and its institutions, including planning law, and then using these to guide societal transformation; or is it the other way round? Is it not a case of the transfer of assets *having* to take place, with planning law merely playing a valorizing and 'protective' function in this process by ensuring that the rights are awarded in the right place at the right time and in the right quantum?

Amidst the many questions and uncertainties seventeen years into the transition, it is clear that without a change in the material conditions, for example the transfer of real/tangible, valuable assets and the capacities to initiate and contribute to economic value-creating processes, and the ability and right to use these assets, real and lasting transformation will not come about. Maybe it is here that planning law could play a role. Planning laws could, notably in urban South Africa, influence the distribution of wealth generated by urban land development, particularly through the promotion of value capture and the facilitation of socially valuable land development on well-located urban land. Planning laws could also play a key role in preserving high urban land values, through constraining the supply of land for development, which serves only to benefit the existing urban land-owners and excludes those who wish to secure a toehold in the urban land market. The promise inherent in this suggestion, for example that land development rights can be created and distributed to a previously excluded group, is, however, very difficult to effect in practice. A clear challenge is the scale of the process in the South African situation: Where the issue is that of a majority responding to the needs of a deprived minority, such as in the more affluent North, the creation of small, carefully planned segments of value in property/land through the allocation of development rights (using planning law) on land located in, or in close proximity to larger areas of high-value land, makes very good sense. Where the complete opposite prevails, and where the objective is to valorize large segments of low-value land amidst a small pool of high-value land, the modalities of doing so are far more complicated and the prospects of achieving the desire outcomes, are far more daunting and far less certain.

A Better Life for All …

While there has been some residential upward mobility and movement into former white suburbs and new golf and security estates, the majority of the Black population still live in peripheral 'townships' and informal areas on the outskirts of towns and cities. While many of the formal 'township areas' have seen the provision of municipal services, the tarring of roads, greening, the provision of new so-called RDP-housing projects, and the construction of shopping malls, these are still, relative to the residential areas in which the majority of more affluent (and still primarily White) South Africans live, low quality human settlements. In addition, Black people living there are still dependent on job opportunities in former 'White areas' and generally lack quality education and the life chances that White youth by and large still enjoy.

During the 1980s and the first half of the 1990s, activists and progressive academics had lively debates on the possible futures of the 'townships'. These debates have now all but disappeared. By default, and largely because of the paucity of alternative views, these township areas are now seemingly set to remain a part of the South African urban fabric and settlement pattern. The State's focus is on *servicing* and making permanent these temporary spaces of servitude created by apartheid, by providing water, electricity, sanitation, storm water and roads, irrespective of the marginality of the settlement. Returning to the earlier discussion on the use of planning law to 'give rights and create value', this can and has happened in many of these townships. This was in fact done by incorporating former 'township areas' into town-planning schemes previously only prepared for 'White areas' in accordance with pre-1994 planning legislation. Giving development rights has, however, not brought with it a resultant increase in property value. It is undoubtedly so that people were living in squalor and that 'servicing' had made life better, but this has essentially been little else than making the poor more comfortable in their misery. The structural and spatial-economic causes of their poverty, for example the location of vast low-quality settlements on the edges of the real economy, have not been changed. At the same time, the use of planning to allocate development rights to properties e with the objective creating land/monetary value in property has not been possible in these areas. The reason is simply that, despite the high need for land, there is a lack of *economic* demand for properties in these areas, irrespective of their zoning.

The More and Less Things Change...

Town planning consultants and officials are by and large today doing what they did in the pre-1994 period. Many of those in this group who earn their living by submitting applications for land development are probably content with the system: They are still able to use the laws they are familiar with to acquire development rights in old and new formal, well-located areas of cities and towns that they are equally familiar with. In addition, and as noted earlier, some consultants seized the option of the alternative route offered by the DFA for submitting applications. By submitting applications to provincial governments in terms of this Act, they were able to sidestep the wishes of municipalities. While this was not the sole reason for using the Act, this practice resulted in often progressive municipal plans for spatial restructuring being outflanked where these stood between planning consultants and the granting of development rights to a developer (see du Toit et al. 2008).

It was noted earlier in this chapter that planning laws were used pre-1994 to ensure stability and protect property values for the White middle and higher income classes; these laws are being used post-1994 to do the same. In this case, the beneficiaries include a new middle and higher income Black group. At the same time, planning laws have enabled the ex-urban sprouting of so-called 'lifestyle estates': gated residential enclaves themed around golf, equestrian pursuits or water-sports, which is the very antithesis of the professed objectives of post-apartheid planning and land development.

Conclusions

Despite the poor showing to date, it may yet be that planning law could play a transformative societal role in this country and it may do likewise in other countries. What the South African story offers with this in mind is some pointers, which also serve as a conclusion to this chapter, as to the conditions that may enable it to play this role.

(a) Champions are indispensable for driving the cause and staying the distance Champions, as agents of change, are required and need to be located both in and outside the State. While the outsiders can 'keep the topic alive', the insiders must craft new legislation and steer it through the minefield of government approval and law-making processes. If such champions can, despite their differences, share a common area of concern and even a professional background, such as Planning, the interaction between those on the inside and outside will be easier. At the same time, as the municipal Integrated Development Plans (IDPs) have demonstrated, the persistence of the department responsible for the Act (that called the IDPs to life) and the plan itself, can be a key driver in municipalities and provinces for complying with progressive legislation, through in this case, the preparation and review of IDPs.

(b) Real change (still) lies in the material world Individual and collective economic and social empowerment is the key driver of transformation. While institutions can regulate, secure and ensure transformation and provide for planning laws that can give and regulate development rights on land, these institutions in the superstructure cannot force the base along. Change can be implemented in the base and must be planned for, as such. The superstructure comes into play once this transformation has been conceived, described and mapped out. From this perspective, planning law would not be the instigator and driver of the transformation process. Instead, it would provide the regulatory framework within which such progressive change would take place, and it would award and regulate property rights. In addition to this role, planning law would sustain the benefits that come from transformative practices and actions by maintaining the regulatory framework, protecting the property values of beneficiaries of land reform, and lowering the land market barriers to accessing urban land. There is a general aspect to this, regarding not just the reconstruction after apartheid, but also the pursuit of more sustainable cities and the transformation that this will require. Planning can channel action and give expression to change initiated by others and in other sectors and spheres of life, and steer it in the right directions, but not conjure up such change out of nothing. At best, it can (1) entice development through incentives and the promise of rights, and (2) steer economic activity, in so doing impacting on the distribution of such activity in a locality or region. In the end though, for those (planners) interested in transformation, there may be more benefit to be had from studying, understanding and changing the profiles of access to, and ownership of resources, and the ways in which a local/regional economies run, than from planning law.

(c) System and agent-recoding and re-engineering is imperative Systems/ Institutions and role-players in systems that have 'crossed a line in history', very easily gravitate back to the place they were, and take on roles similar to the ones they had in the previous dispensation. In order to counter this institutional lock-in, the DNA of the system and its agents needs to be changed. While this does not suggest that every planner must and can be re-engineered to become a reborn visionary, there should be at least a generally shared concern amongst planners with 'the future' and an appreciation of how the move to more desirable futures can be supported (and frustrated, and this countered) in any planning workplace/position and task.

(d) Recognizing power and resisting the allure of simplistic solutions Government systems based on deliberation and debate tend to lose out against control-centered ones. In the case of South Africa with its deliberative, cooperative governance model and the densely interwoven intergovernmental engagements in the domain of planning, there (unfortunately) seems to be a growing allure for simpler, cruder, more control-based systems. Rights-driven conceptions of planning law and societal transformation, even if they are grounded in and entail high levels of deliberation, can be hijacked by those in the control-camp in their pursuit of quick-fixes. Under such conditions, development rights could be allocated to properties

without recognizing that societal transformation is an evolutionary, deliberative process located in a contested, intertwined and messy world of *inter alia* planning law, property rights and planning ideals.

(e) An appreciation for scale, complexity and time Transformation, even in small towns is a massive, and a constantly changing endeavour. Simplistic symmetric views, such as that 'planning law broke it, planning law will fix it', and static views about 'what is' and 'where we are going', are not useful. Likewise, failure to appreciate the challenges associated with transformation, such as 'new elite establishment' and the abuse of (new and old) instruments to protect newly acquired wealth and privilege, need to be anticipated and proactive preventative measures taken. Not recognizing this, and not planning to stay the distance, means that the desired transformation will not take place.

(f) The influence of external discourses The introduction of international discourses and wisdoms, in this case those around decentralized developmental planning, may not be without merit, but the under-considered imposition of such models on ill-prepared local governments struggling to find their feet, may be. The same goes for the suggestion that imported planning law regimes that have served formal settlement development and regulation processes, and that have allocated development rights and created and protected land values in such settlements, could be expanded to achieve similar results in less formal settings and settlements. The prospects for ensuring progressive outcomes through the deployment of such rights-granting regimes in informal situations deserve more exploration. The South African society, still young and in the institutional and societal making, can serve as one such area for further research as it and its institutions mature.

(g) Planning law and Government objectives and policy Planning law is an instrument that takes its cue from State policy. With this in mind, what is required for the State in South Africa is to (1) decide on the vision, set the development objectives and put in place the directive and enabling policy framework to take the country there; (2) promulgate a planning law that can entice, enable, regulate and steer land development in pursuit of these objectives; and (3) implement and apply the planning law, using all the necessary instruments it has to enable and enhance this action. History has taught that this is not an easy task; hence the need to start right away.

References

African National Congress 1994. *The Reconstruction and Development Programme*. Johannesburg: Umanyono Publishers.

Berrisford, S. 2011. Unravelling apartheid spatial planning legislation in South Africa. *Urban Forum*, 22(3), 209–228.

Development and Planning Commission 1999. *Green Paper on Development and Planning*. Pretoria: Government Printer.

Du Toit, J, Oranje, M. and Landman, K. 2008. *Beyond Policy Frameworks: Listening to the Voices of Developers and Municipalities in Urban Land Development in South Africa*. Technical report. Pretoria: Urban LandMark.

Mabin, A. 1991. A quarter century of urbanisation: 1965–1990. *Town and Regional Planning*, 31, 17–25.

Mabin, A. and Smit, D. 1992. *Reconstructing South Africa's Cities 1900–2000: A Prospectus (or a cautionary tale)*. Paper presented at the African Studies Seminar. Johannesburg: University of the Witwatersrand, 18 May 1992.

Muller, J. 1983. *Theory and Practice: The Janus Face of Planning*. Inaugural lecture presented on the 2nd of September 1982. Johannesburg: Witwatersrand Press.

Muller, J. 2003. *Thought and Action: A Personal and Partial History of the South African Planning Profession*. Centrahill: QC Publications.

Oranje, M. 1998. *The Language Game of South African Urban and Regional Planning: A Cognitive Mapping from the Past into the Future*. Unpublished PhD thesis submitted in partial fulfilment of the requirements for the degree PhD in Town and Regional Planning at the University of Pretoria. Pretoria, South Africa.

Oranje, M. 2002. Should provinces/regions have their own planning acts? An exploration of the debate using the post-1994 South African experience, in Y. Rydin and A. Thornley (eds), *Planning in a Globalised World*. Aldershot: Avebury, 83–116.

Oranje, M. 2011. Neo-liberalism, shallow dreaming and the unyielding apartheid city, in G. Baeten and T. Tasan-Kok (eds), *Contradictions of Neo-liberalising Planning: Cities, Policies and Politics*. Amsterdam: Springer.

Parnell, S. 1993. Creating racial privilege: the origins of South African public health and town planning legislation. *Journal of Southern African Studies*, 19(3), 471–488.

Republic of South Africa 1996. *Local Government Transition Act: Second Amendment Act, No 97 of 1996*. Cape Town: Government Printer.

Chapter 4
Talking about Property Rights over Tea: Discourse and Policy in the US and Europe

Harvey M. Jacobs

A radical right has emerged in both the US and in Western Europe. In both regions their discourses and policy agendas are largely the same: smaller government, less regulation, lower taxes, flexible labour markets, job creation, strong borders and anti-immigration. This rhetoric appears to appeal to an increasingly large segment of voters. In the US it has led to the so-called Tea Party Movement and in Europe to the election of right-wing candidates and parties in traditionally social democratic countries such as Sweden and Finland.

A key component of traditional rightist rhetoric is about the foundational basis of land-based property rights to assure political freedoms and the functionality of markets. This argument is rooted in classic political-economy texts from the seventeenth and eighteenth centuries. This rhetoric has the potential to constrain public sector spatial planning – land use, environmental and redevelopment planning and regulatory activities.

This chapter examines the way in which property rights is a part of contemporary rightist discourse in the US and Europe and why similarities and differences exist. Speculation is then offered about what role property rights will play in the future, and how this is likely to affect the opportunity and ability to engage in public sector spatial planning.

A Global Context for Talking about Property Rights

Private property is a social and legal institution with a long history (Schlatter 1951). It has come into contemporary focus because of the changing nature of the global political economy.

With the fall of the Berlin Wall in the late 1980s and the dissolution of the Soviet Union in the early 1990s, some commentators believed that the grand social debates of the twentieth century were finished (Fukuyama 1989). Throughout the century, the debates had been structured by the relative merits of conflicting political economies: socialism versus capitalism and communism versus democracy. In the new era, it seemed that only one set of ideas would prevail:

capitalism and democracy.[1] The new countries of Central and Eastern Europe and the former Soviet Union, as well as other countries that were undergoing their own independent political changes (such as South Africa) began asking themselves and others how to become more integrated into the global community. How does a country acquire the economic standing of the advanced developed countries? How does a country acquire the political legitimacy of the advanced developed countries? How does a country acquire capitalism and democracy? These became among the most pressing sets of questions of the late twentieth century.

An answer centred on private property. Private property was offered as the literal key to a market-based capitalist economy; likewise, private property was central to democratic political structures.

Over the last two decades, developing and transition countries around the world have, with the counsel of the multilateral and bilateral international aid agencies, moved to introduce the social and legal institutions of private property (for example Alexander 2006). This tendency has been further aided by advocacy suggesting that the creation of private property is the central variable to alleviation of poverty in developing countries (De Soto 2000). Some suggest that the extent of private property rights serves as a reliable indicator of both economic strength and political freedom, leading to global rankings of private property rights robustness (Jackson 2011).

As the global discussion about private property rights has accelerated, one focus has become the status of property rights in the developed countries themselves. My purpose here is to understand the changing discourse about property rights and its likely impact on spatial planning. Prior research suggests that such advocacy has been influential, in both the US and Europe (for example Jacobs 2003, 2008a, 2009).

The American Tea Party (and its European Counterparts)

The American Tea Party movement burst into politics in 2009. They are broadly recognized as a conservative movement that promotes an agenda focused on reduced government spending, opposition to taxation, reducing the federal deficit and adhering to an originalist interpretation of the US Constitution (Wikipedia 2011b).

In a recent essay about the implications of the Tea Party for American foreign policy, Mead (2011: 32) notes that while 'there are reasons to doubt the movement's long-term ability to dominate politics across the rest of the country … the Tea Party movement has clearly struck a nerve in American politics'. Mead (2011) presents the Tea Party as the latest manifestation of a form of American populism called Jacksonianism (after the 1830s American president, Andrew Jackson).

1 Huntington's (1997) notion of a 'clash of civilizations' is an alternate concept to the one advanced by Fukuyama (1989).

Philosophically he ties them to the Scottish Enlightenment – the idea that the average person (in contrast to elites) can discern the truth in public, religious and scientific affairs.

Who are the adherents of this movement? A 2010 national poll found them to be older, more likely to be men, better educated and with higher incomes than the average American. They describe themselves as politically conservative and they attend religious services more regularly than the average American; more than half of them keep a gun in their home. Overall they are both more pessimistic about America's future and angrier than the average American about the current state of public policy, the economy and social affairs. A very large majority of Tea Party supporters believe that their views reflect the views of most Americans, though this does not appear to be the case (Montopoli 2010).

While the Tea Party is largely focused on economic issues, they do pay attention to some of the social issues long associated with the American political right: illegal immigration, abortion, same-sex marriage and gun control. In addition, they are strongly opposed to the healthcare law adopted under President Obama's leadership. Two other prominent issues for them are global warming and cap and trade. Tea Party adherents doubt the existence of global warming and oppose the implementation of an emission trading system to manage air pollution (Montopoli 2010; Wikipedia 2011b).

Mead notes that the populist tendency represented by the Tea Party has always had a land-based element associated with it

> [T]he demand of Jacksonian America for cheap or, better, free land in the nineteenth century led to the Homestead Act, which allowed millions of immigrants and urban workers to start family farms. It also led to the systematic and sometimes genocidal removal of Native Americans from their traditional hunting grounds and a massively subsidized 'farm bubble' that helped bring about the Great Depression. Populist hunger for land in the twentieth century paved the way for an era of federally subsidized home mortgages and the devastating burst of the housing bubble. (2011: 33)

While Mead begins his examination wondering about 'the movement's long-term ability to dominate politics across the rest of the country' (2011: 32), he ends on a more sombre note. 'If the Tea Party movement fades away, other voices of populist protest will take its place. American policymakers … cannot do their jobs well without a deep understanding of what is one of the principal forces in American political life' (Mead 2011: 44).

The situation in Europe is similar in that there has been a parallel rise of conservative political forces, and these forces are described and understood as having a populist basis. The National Front in France, Lega Nord in Italy, the FPÖ (Freiheitliche Partei Österreichs – Freedom Party of Austria) and Jörg Haider in Austria are examples that have been studied and broadly discussed in the academic and popular literature (see, respectively, Mayer 2003, Woods 2009 and Höbelt 2003; and more generally Schain et al. 2002 and Ignazi 2003). Added to this are the more recent success of nationalistic conservatives in Sweden and Finland in the 2010 and 2011 elections, the rebirth of the National Front in France under Marine Le Pen, and the expected success of other, similar parties throughout northern Europe (Wikipedia 2011a; *The Economist* 2011b, 2011a; Shorto 2011). In all these cases, some combination of the substantive issues parallel to those fuelling the American Tea Party have been at the root of the conservative turn – a mix of social and fiscal concerns, for example about immigration, citizenship, government spending and debt.[2]

Discourse and Policy in the US

Discourse about property rights is long-standing in the US. The role and place of property rights was of intense interest and debate among the founders. For a variety of reasons – philosophical, historical and contemporary – there was a strong sense that the right to hold and control property was an important element of a democratic governmental structure.

First there was the reality of the settlement process. Colonial America was settled by Europeans searching for religious and political freedom (the rights guaranteed in the First Amendment of the Constitution's Bill of Rights) and for access to land (Ely 1992). In Europe in this period an elite owned most of the land, and the prospects for the ordinary person to obtain freehold ownership was small. America offered an alternative. America was a place where any white male immigrant could, in theory, get ownership of land, and with that land as capital make a future for themselves. America was, quite literally, the land of opportunity.

During America's colonial era John Locke influenced the founders' ideas about ownership and democracy. One came to possess property through using it (which provided the justification for taking land from America's native inhabitants, who were not using it in the European sense of active agricultural and forest management), and freely constituted governments (for example democracies) existed for the protection of individual liberties, including the liberty to hold and control property.

The country's founders configured these ideas into a particular and specific relationship. In the debate over the ratification of the proposed US Constitution, James Madison wrote in *Federalist* No. 54 in 1788, 'government is instituted no less for the protection of property than of the persons of individuals' (Hamilton, Madison and Jay 1961: 339). Others, including Alexander Hamilton and John Adams concurred. Adams (1851 [1790]: 280) noted that 'property must be secured or liberty cannot exist. The moment the idea is admitted into society that property

2 Depending on the country and the situation particular issues have more or less prominent.

is not as sacred as the laws of God, and that there is not a force of law and public justice to protect it, anarchy and tyranny commence'.

Thomas Jefferson left modern Americans with an enduring image of this perspective – that of the yeoman farmer. According to Jefferson, because the yeoman farmer owned his own farm, and could produce food and fuel for himself and his family, he was obligated to no one – he was literally free to exercise his political views as a democrat. For Jefferson it was the very act of ownership that created the conditions that allowed democracy to exist.

But this view of the relationship of property to democracy was not unchallenged. Also drawing from Locke, other founders saw the need for private property ownership to bow to social needs. Benjamin Franklin was perhaps the most articulate proponent of a counter-position to that of Madison, Adams, Hamilton and others. For example, in the debate over the ratification of the Pennsylvania state constitution, Franklin (1907 [1789]: 59) said: 'Private property is a creature of society, and is subject to the calls of the society whenever its necessities require it, even to the last farthing'. Franklin did not see property rights as sacrosanct.[3] He viewed as legitimate the public's right to create, re-create, take away and regulate property as it best served public purposes.

Private property was thus a confusing issue for the founders. How were these disparate positions resolved? With ambiguity. In 1776 the Declaration of Independence promised each (free, white, male) American 'life, liberty and the pursuit of happiness'. What is striking about this phrase is that Thomas Jefferson, the Declaration's author, borrowed it from Locke. Locke's phrase was life, liberty and property. This is what Jefferson wanted the Declaration to say, but he did not hold sway.

In 1787 the US Constitution was adopted. What did it say about land-based private property? Nothing![4] It was not until 1791 with the adoption of the Bill of Rights that the now infamous and contentious so-called 'takings' phrase appeared as the closing clause to the Fifth Amendment to the Constitution: 'nor shall private property be taken for public use, without just compensation'.

3 These sentiments by Franklin were not isolated. As noted by Brands (2000: 623) 'Franklin took a striking socialistic view of property.' Brands (2000: 623) provides these other examples of Franklin's opinions: 'All property … seem to me to be the creature of public convention … All the property that is necessary to a man for the conservation of the individual and propagation of the species is his natural right, … but all property superfluous to such purposes is the property of the public, who by their laws, have created it, and who may therefore by other laws dispose of it whenever the welfare of the public shall demand such disposition.'

4 The US Constitution does speak to private property, but just not about land-based private property. What the Constitution recognizes are slaves as property under Section 2 of Article IV, where it establishes the right of owners to have escaped slaves returned to them. Also under Section 2 of Article III, the Constitution establishes a procedure for how conflicting claims to state-based land grants by individuals would be resolved.

With the adoption of this phrase, the Constitution formally recognized the existence of private property, an action denoted as taken (commonly denoted as takings), a realm of activity which is public use, and a form of payment specified as just compensation. Where private property exists, it may be taken (for example seized by the government without the landowner's permission) but only for a denoted public use, and when just compensation is provided. If any of these conditions are not met, then a taking may not occur. But the clause does not say, and colonial commentary does not clarify, what constitutes private property, exactly when a taking has occurred, what is a public use, and what makes up just compensation.

For over 100 years the place of private property in American democracy and the exact meaning of the takings clause were largely theoretical. There was little regulation of land as we currently understand it. And when government determined that it needed to take property, the public use was generally clear – land for a school, a road, or other public facility – and the owner was compensated. For much of the eighteenth and nineteenth centuries there was little public discourse or social conflict over private property. The US had land in abundance, and it was the disposition of public land, not the status of private land, that dominated the policy agenda (Gates 1968).

The twentieth century ushered in a new period for American land policy, and thus discourses over property rights. The frontier was settled (Turner 1893). Public policy shifted from the disposition of America's public lands to the management of its land resources. The turn of the century (1880–1920) was a period of intensive immigration, industrialization and urbanization. The 1920 US Census officially recorded the shift from a rural to an urban nation. In response to these changes, modern spatial planning and policy and the modern relationship of the state to the individual was born. Cities and states began to pass varied regulations to manage public health and safety conditions, and out of these new conditions arose a concern about the appropriate limits to government regulation.

The US Supreme Court found itself called upon to clarify the place of private property in modern America.[5] The Court was asked whether there were limits to public regulation of private property. Their initial answer – no. In 1915 the Court concluded:

> It is to be remembered that we are dealing with one of the most essential powers of government, one that is the least limitable … the imperative necessity for its existence precludes any limitation upon it. … There must be progress and if in its march private interests are in the way they must yield to the good of the community. (*Hadacheck v. Sebastian* 239 US 394 (1915): 410)

5 There is a huge body of scholarship about how to understand and approach the jurisprudence of the court. For the purposes of this discussion, I rely heavily on the analysis and interpretation of Kayden (2004).

But conditions of the period kept the issue before the Court. In *Pennsylvania Coal v. Mahon* in 1922 (260 US 393), the Court appears to completely change its mind about the reach of governmental power. It was here that the Court issued its famous dictum defining the twentieth century concept of regulatory taking: 'The general rule ... is, that while property may be regulated to a certain extent, if regulation *goes too far* it will be recognized as a taking' (260 US 393 (1922): 415; emphasis added). A regulation could be equivalent to a takings under the Fifth Amendment, and if it was then compensation was required. This 'warning' has echoed down through the century. Its impact has been to make planners, but even more so the locally elected officials who hold authority for formally adopting and implementing spatial plans, cautious in the planning process. If a spatial plan can be equivalent to a physical expropriation, government might find itself liable for compensation. Given the decentralized structure of US governance, this would put the burden on the most local of governmental units that are given the authority to develop and implement spatial plans.

The final case of importance from this period was the Court's ruling on the validity of zoning. New York City is credited with inventing zoning in 1916. Within a few years zoning had spread across the country as a way for cities to manage growing populations, industrialization, and property values (Revell 1999; Wolf 2008). In 1926 the Court examined whether the idea of allowing a local government to regulate land use by designating land-use zones, which provided for different levels of development opportunities, was acceptable (*Euclid v. Ambler Realty* 272 US 365). The Court decided yes, such an approach to the management of private property rights was acceptable.

So as the Depression loomed, the Court had said: there are no limits on public regulation; yes there are limits, regulation that 'goes too far' is unacceptable; but regulation of private property through zoning is not going too far. So where was 'too far'? The Court would not define this in advance. In practice this was not a problem, as most governmental bodies did not use their authority to impose onerous requirements upon landowners.

After *Euclid*, the Court largely left the property rights arena for fifty years. In the late 1970s they re-entered it with vigour, seeking to establish boundaries to governmental authority. Many things had changed, including America's attention to private property. Starting around 1970 the US experienced an explosion of laws, policies and regulations at the national, state and local levels that affected private property. The Clean Air Act, Clean Water Act, Coastal Zone Management Act, National Environmental Policy Act are all examples at the national level (Moss 1977 provides one then-contemporary chronicling of these). Among state governments, a so-called 'quiet revolution in land-use control' had occurred where nearly a dozen states re-asserted their Constitutional authority to regulate private land-use activities at the state level (Bosselman and Callies 1971). And local governments across the country were beginning what have become decades-long experiments in public policy approaches to protect and manage farmlands, wetlands, open spaces, watersheds, threatened habitats, urban sprawl, etc.[6]

Between 1978 and 1994 the Court heard a series of cases in which they worked to more precisely define the rules of interaction between government and private property owners.[7] While the Court did not eviscerate the right of government to regulate private property, the Court did begin to more clearly say when the line of 'too far' articulated in *Penn Coal* had been crossed. Of these cases one that received a great deal of attention was *Lucas v. South Carolina Coastal Council*, 505 US 1003 (1992). Here the Court ruled that when all economically viable use has been taken by regulation, this was an instance of regulation going 'too far' and compensation was owed the landowner.[8] In general, regulation and spatial planning were still acceptable, but a regulating body needed to be careful in its formulation and administration.

The twenty-first century began with two major cases being decided by the Court, both of which seem to take a step back from the boundary-setting tone of the immediate prior period. In *Tahoe-Sierra Preservation Council v. Tahoe Regional Planning Agency* (535 US 302 (2002)), the Court took up the matter of a nearly three year moratoria on development. In a decision strongly in favour of government, the Court found that local planning and regulation are normal and expected governmental functions and that the Court had no reason to interfere with them (Kayden 2002).

Then, in June 2005, the Court issued its closely watched decision in the case of *Kelo v. City of New London* (545 US 469 (2005)). In the *Kelo* case the city asserted its right to take private property, with compensation, for a public use, when the public use was defined to be consolidation of land for distribution to another private owner in order to facilitate economic development in the city. Of a total of 115 properties, fifteen property owners challenged the legitimacy of the proposed taking, arguing that expropriation for economic development purposes violated the intent of the takings clause. It was further argued that if the city prevailed, government could always assert that a proposed new use was in the greater public purpose (Jacobs and Bassett 2011).

---

6 Daniels and Bowers (1997) is an example of these approaches applied to farmland protection.

7 Some of the most prominent and discussed examples include the decisions of the Court in the cases of *Penn Central Transport. Co. v. New York City* 438 US 104 (1978); *First English Evangelical Lutheran Church v. County of Los Angeles*, 482 US 304 (1987); *Nollan v. California Coastal Commission*, 483 US 825 (1987); *Lucas v. South Carolina Coastal Council*, 505 US 1003 (1992); and *Dolan v. City of Tigard*, 512 US 374 (1994).

8 However, even in Lucas the Court provided government with an 'out'. That is, the Court noted that its ruling was made with the proviso that the individual's use of land could not violate state-based background principles of nuisance which governed property use (505 US 1003 (1992): 1029).

By a one vote margin the Court sided with the city and against Kelo. The decision was one of the Court's most controversial of the first decade of this century. And the Court itself was aware that it would be. According to the Court, their decision was only about whether New London's action was acceptable under the national Constitution. They then went out of their way to invite state-based action in response to the decision by saying 'We emphasize that nothing in our opinion precludes any State from placing further restrictions on its exercise of the takings power' (545 US 469 (2005) at 489).

Parallel to the rise of post 1970 land use and environmental regulation came the formation of the so-called property rights movement (Gottlieb 1989; Miniter 1994). From their perspective the intent of the key American founders and of the principles embodied in the founding documents are clear – the protection of private property rights is to be a key element of the American political and social contract (Jacobs 1995, 1998b; Marzulla 1996). They hold an alarmist view of twentieth century public policy and law. The century presents a story that moves away from a view of property rights as central to liberty and democracy. Instead, what appears is a story in which government is allowed ever increasing authority to reshape property without respecting the protections afforded to property by the Constitution. Despite the promise contained in *Penn Coal* (1922), the right of government over the rights of individuals seems to be continually affirmed (Bosselman et al. 1973; Salkin 2001). Even the Court's decisions of the late 1980s and early 1990s appeared to have had little impact on governmental practices (Roddewig and Duerksen 1989).

The property rights movement has pursued a multi-level strategy to achieve their objectives (Jacobs 1999).[9] For many reasons they have focused their efforts at the state level, and here they have found fertile ground for their arguments and ideas.[10] Since 1991 every state in the US has considered state-based legislation in support of the policy position of the property rights movement, and by the decade's end 27 states have passed such legislation (Emerson and Wise 1997, Jacobs 1998a, 1999b). These states are on both sides of the Mississippi, they are 'red' and 'blue' states, and extend from Maine to Washington, the Dakotas to Texas, with eleven of these states east of the Mississippi River.[11]

9 There have been and are a wide range of advocacy groups pursuing aspects of this strategy. These include, for example, the American Land Rights Association, the Competitive Enterprise Institute, the Castle Coalition. These groups can be searched on the web, and they in turn have links to related organizations.

10 The states have always been central players in the land-use drama. It is they who have residual authority under Article X of the Bill of Rights for land use, which in turn leads them to adopt enabling legislation passing this authority along to sub-state governments.

11 The use of these color designations came into being after the Bush-Gore presidential election of 2000. The media depicted states where a majority had voted for Bush as red states and where a majority had voted for Gore as blue states. Subsequently, this designation came to be used as shorthand in American political discourse for areas of the

From 2005 forward the *Kelo* case has been used by the property rights movement to invigorate their agenda and spur public discourse. The property rights movement wants citizens to talk about the issues of when it is reasonable and legitimate for government to take property under the authority of the takings clause, and whether there are limits to reasonable government regulation, beyond which the individual property owner is entitled to a degree of compensation (á la *Penn Coal*).

The property right movement's success in bringing this conversation into the public realm has been breath-taking. They have managed to put the issue into the most mainstream of media fora, and following on the Court's invitation for state-based action, they have orchestrated adoption of anti-*Kelo* laws in 43 of the 50 US states (Jacobs and Bassett 2010, 2011). The explicit intent of most of these laws is to prohibit eminent domain actions for the sole purpose of economic development, and where privately owned land is taken from one owner to be redistributed to another owner. *Kelo* has allowed the property rights movement to give their issue and perspective public visibility, to make it a national issue, and, importantly, to establish the issue as one of oppressive government and vulturous big corporations versus the common person.

Notwithstanding the actual impact of the state laws adopted, the most significant outcome of the property rights movement's activism has been in the area of public discourse (see Jacobs and Bassett 2010 and 2011; Lopez et al. 2009; Morris 2009; and Somin 2009 for discussions of the impact of these state laws). According to Ely

> the most significant impact of *Kelo* may well be heightened public awareness of the need to guard property rights … [a] welcome if unintended consequence of *Kelo* has been to restore the rights of property owners to public dialogue. (2009: 150)

Nadler et al. agree:

> the *Kelo* decision seems to have tapped into existing concerns about the sanctity of the home, government overreaching, and tensions between protecting public goods (like the environment) and protecting private rights. In some sense, *Kelo* was a "perfect storm" because all these issues were directly implicated in the decision. (2008: 306)

Thus in the United States there is a paradox about property rights. Legally the courts largely affirm the right of government to manage (regulate) these rights to achieve public purposes. Administratively, most governmental units do not act to impose upon those rights in a particularly stringent way. Yet there is renewed

country which appear to express more conservative (red state) or liberal (blue state) social and political values; see, for example, Brooks (2001).

interest in property rights law and policy as a conscious part of a multi-decade strategy from the political right. Over the last thirty years the so-called private property rights movement has grown substantially. There are now a wide range of advocacy groups pursuing aspects of an integrated policy agenda, as well as a supporting network of legal advocacy firms available to bring action on behalf of landowners (such as Susette Kelo) and against elected officials (Teles 2008). By some measures, including media attention and state laws passed, these groups have been wildly successful (Sagalyn 2008 for one discussion of the media impact).

This paradox directly impacts the domain of spatial planning. In theory there are few practical limits to the spatial plans that planners can propose and governments can adopt and implement. But in practice the rhetorical focus on property rights, the media success of the property rights movement, and the concern about crossing the largely undefined line of what *Penn Coal* identified as 'too far' means that spatial planners are cautious in what they propose, and elected officials are even more cautious in what they choose to adopt. So spatial planning exists in a realm where its opponent has little formal legal basis for challenging any planning, but the discourse about property rights has created a climate which impedes the scope of the spatial planning that occurs.

Discourse and Policy in (Western) Europe

The situation in Europe is similar and different. As in the United States, throughout Europe there is a network of politically conservative, policy advocacy organizations. Broadly speaking they advocate less government regulation, stronger markets, and stronger individual liberties – a liberal position in the European sense of the term. But their focus on the discourse about property rights seems to be much less than their US counterparts.

The Stockholm Network functions as a coordinating and networking body for over 120 of such groups throughout Europe (http://www.stockholm-network.org/). Their website groups activities into three broad areas: health and welfare reform, energy and environment (with further focus on climate change and cap and trade proposals), and intellectual property and property rights. Noticeably absent is the type of land-based property rights advocacy common in the US. The Centre for a New Europe, formerly headquartered in Brussels, was similar in focus (the Centre appears to have closed in 2008 or 2009). When functioning, it acknowledged on its 'about CNE' page the key role of property rights: 'Private property ensures the freedom of the individual in his private sphere' (http://www.cne.org/about2.htm). Many of these groups come together for a networking conference known as the European Resource Bank (http://www.resourcebank.eu/european-resource-bank-meeting-2011/about.html). According to its website the bank is 'modelled after the resource bank organized every year by the Heritage Foundation in the United States … [and is] the largest annual congress of free market think-tanks in Europe'. They have been meeting annually since 2004. As is true with the

Stockholm Network, the 2011 conference showed little evidence of interest in land-based property rights per se (http://www.resourcebank.eu/european-resource-bank-meeting-2011/agenda.html).

A notable exception is the International Centre for Research on Environmental Issues in France (http://www.icrei.org/). ICREI was founded in 1992 with an explicit mission to undertake research and education promoting 'property rights and market transactions, assessing the net impact of social regulations, using the principle of responsibility and promoting institutional arrangements for a free society'. Since 1996 they have organized a biennial international conference with a single environmental resource focus – theory and ethics (1996), water (1998), marine resources (2000), coastal zone (2002), waste (2004), land resources (2006), air pollution and climate change (2008), biodiversity (2010) and forthcoming in 2012, agriculture and forestry – from 'the perspective of non-regulatory tools, for example property rights and markets'.

Even in those areas where they actively seek influence, it is unclear how much impact liberal advocacy organizations have had (though the 2010 and 2011 elections in Sweden and Finland and forthcoming elections in France give pause to this assertion). Regardless, it does not appear that, with the exception of ICREI, the European groups have chosen to focus on the property rights debate in the way the sub-set of American groups have. But this is not the end of the story. There are two other elements to be explored that give a fuller picture of public discourse in Europe – public attitudes and legal decisions.

In 2006 I undertook case study research in Europe (France, Italy, Netherlands and Norway) about the status of property rights. Interviews with local and regional officials and advocacy group representatives suggest a significantly changed mood. Here I examine two of these cases – southern France and Norway.

France presents a perplexing case with regard to the status and future of private property rights and land development. Formally – by statute and under the *Code Civil* – owners enjoy a robust set of rights (Dubois 1998). The right of ownership is set within a legal-administrative structure where there is a tradition of both a strong central state, with significant direct presence in the regions, and a highly fragmented and strong system of local government. As part of this structure, French law allows public sector planners to designate land for non-development, the owner of that land has no claim for regulatory compensation following from this designation, and public sector planners may utilize the power of pre-emption to intervene in proposed individual land sales (Alterman 1997; Dubois 1998).

France's pre-emption power may be as strong as any in Europe. This puts local landowners in a very difficult situation. To the extent land exists within an area designated as subject to pre-emption, local landowners may not engage in market-based sales prior to a declaration filed with the government. Failure to make this declaration voids all proposed sales. Once a declaration is made, however, the eventual outcome of the land transaction is very much controlled by the public sector. Assuming the public sector wants to pre-empt land, landowners have several choices: to accept the offer of sale to the public sector at a price determined by the public, or to not accept the offer from the public but to then to continue to hold on to the ownership of the land. The landowner does not have the right to exercise a market sale solely on the basis of not liking the price offered by the public sector.

This situation is dissonant with a period from the 1980s onward of strong, even raging, urban growth (especially in the secondary, provincial cities), where the local governments often lack the skill and resources to respond to this growth. One of the reasons they have difficulty responding is the 'real politic' of local governance – 'mayors feel reluctant to offend local landowners, to whom by virtue of the small size of most communities they would inevitably be very close' (Booth 2003: 954).

In the Montpellier and Nimes regions all interviewees noted the pressures brought by landowners upon local officials to designate their land for development permission. This was especially true of peri-urban agricultural land owners, who are caught in a squeeze between the greater public interest in retaining their land in agriculture, European Union (EU) policies which are pushing for a reduction in agricultural production, and the wide disparity (a scale of ten times or more) between land prices for land use in agriculture and its use for development.

What does all this mean for the status of property rights? It is not clear. Despite the apparent strength of the French public sector, interviewees asserted the apparent rise of landowners' position vis-à-vis the state. So Gambier (2006) argued that overall 'the rights of owners are getting stronger'. Why? He argued that it was because of the rhetoric coming out of Brussels (the EU). That is, it was not policy per se that was strengthening the rights of landowners but a social attitude that they are tuning in to, an attitude that is reflective of new ideas: in his words 'the liberal way of thinking'. Specifically, Gambier noted that in his area of expertise – the exercise of the pre-emption rights by the public sector – 'judges are (increasingly) backing the private (owner) against the public'. Girault (2006) followed in this line of thinking. With regard to the use of the pre-emption right, she argued that if a proposed pre-emption is challenged by an owner, the 'court would back the owner, unless it is a "real" project' (for example a road, a school, etc.). She attributed the rise of a property rights attitude to the culture and traditional attitude toward land and environmental resources. So, 'local people here are fed up with protection' because of all the rules that exist, often EU rules. 'People don't want to be imposed (upon); they are more liberal.'

In France urban and peri-urban growth pressures have been increasing for decades. Landowners have strong motivations to seek options for their land outside of agriculture. As Bourdat (2006) noted, 'a lot of vineyards are disappearing'; as a result 'owners put lots of pressure on local communities so that their land can be developed. It is difficult not to listen to them', especially because 'we don't have easy solutions for them'.

In Norway private property has a long and strong place; interviewees noted that it is an idea that dates at least to the Viking era (Rasmussen 2006). In the 20th century, the Norwegian Labour Party formally adopted the idea of private property ownership by all working Norwegians as a key component of their party platform (Jensen 2006).

Norway, like France, has strong public controls over land and especially over the process of rural land subdivision and sale. While '[i]n principle land is freely disposable', in reality '[m]ost acquisitions of property in Norway are subject to the Concession Act [of] 1974 ... Under the 1974 Act, all property acquisitions require prior approval from the authorities.'[12] 'If a transaction requires approval ... the government and, more rarely, the local municipality, will have a right of pre-emption' (Askheim and Rødsæther 1998: 425). According to Aasen (2006) there is strong power for the government to say 'no' to proposed agricultural land development (subdivision).

But interviewees spoke of a set of tumultuous changes in Norwegian attitudes toward private property, and as a result changes in the relationships between the individual and state over property. As in many countries, use options available to agricultural land owners were among the most controversial of issues. Falleth (2006) says that the biggest problem in rural Norway today is that 'farmers want to build infrastructure (to accommodate opportunities) for tourism, with amenities. So there is a conflict between agriculture and (proposed) urban (style) uses.' The Norwegian Mountain Touring Organisation is especially concerned about this.

> There are two different pressures on land: 1) a lot of Norwegian people want their own cabin in the mountains; rural landowners can earn money from sale of their land for this purpose; 2) the current state of Norwegian agriculture, especially the support from the state; this support is declining. We believe this support will continue to decline. (Hjelle 2006)

This sets up a conflict.

> Landowners say: how will we earn money? The government says – tourism. The question becomes what kind of tourism; how to use nature. A lot of these conflicts with farmers and landowners wanting to earn money come from the government saying you have to earn money from tourism. (Hjelle 2006)

'Now [in rural areas] the focus is on second homes, golf courses, fishing rights; it all has a money focus' (Rasmussen 2006). Rasmussen went on to note that 'the Ministry of Agriculture is putting pressure on the Ministry of Local Government to define nature-tourism as a form of agriculture'.

Part of the reason for this new area of conflict is the introduction of classical liberal ideas into political and policy discourse. Jensen (2006) notes that 'everything was opened to the market in the 1985–1989 period; this was a major

---

12 'The main exceptions include the acquisition of small residential property ... land and buildings not exceeding 5,000 square meters; and transactions between members of the same family' (Askheim and Rødsæther 1998: 425).

shift in Norwegian society. As liberal thinking has come up, those in charge of the public interest are more like businessmen.' Jensen sees several consequences of these changes: 'a loss of trust in society taking care of you and being fair, and (the fact that) you want choices (like in a supermarket)'. Others agree. 'This is a completely different country than it used to be since the late 1980s with the introduction of neo-liberal ideology' (Tesli 2006). One consequence of these changes has to do with the perceived legitimacy of the state to manage property. According to Sevetdal (2006) 'the justification of strong regulation by government is losing to the interests of property, all kinds of property – public, semi-public and private'. Or as a younger, state bureaucrat said, today 'individual rights have a much stronger focus in Norway ... than they have in earlier modern times' (Rasmussen 2006).

Thus Norway, like France, presents a perplexing case. At the surface, based on formal law and administrative practice, it is a country where the state holds a strong position vis-à-vis the individual with regard to property rights. While the individual is said to have a full bundle of rights, and there is a millennial-long respect for private property, tradition, history, and law are all construed to put the central state in a strong position. But there is reported to be an ever louder call by landowners to respect their rights in land and to allow them to do with their land what makes individual economic sense. Bureaucrats, specialists, activists and academics believe there is a fundamental change underway in Norway. Those interviewed shared a sentiment that the property rights of individual owners were getting stronger relative to the position which such rights had held in the past. The forces that are pushing these changes seem to be the same ones that are facilitating change in southern France.

Parallel to and reinforcing these social changes are legal changes.[13] The European Convention for the Protection of Human Rights and Fundamental Freedoms (commonly known as the European Convention on Human Rights) was signed in Rome and adopted in November 1950. Its signatories included the then ten countries of the Council of Europe. A European Court of Human Rights was created to supervise compliance with the Convention. Since its adoption,

13 The US and most of continental Europe function under two different legal systems. The United States, as a former British colony, is based on a common law system, whereas continental Europe, operates under a civil law system, and this system is strongly derived from the Napoleonic Code of 1804 (Booth (2005) discusses these two systems for Britain and France). (There are other ways of organizing European legal and administrative systems; Newman and Thornley (1996) present a five part taxonomy.) In theory, a civil law system operates in such a way as to emphasize the centrality of parliament-based statutes and laws over judge-made decisions. Judicial decisions, specifically precedent as established by prior decisions, do not have the strong role that they do in common law systems. And in fact, drawing from the experience of the French Revolution, civil law systems have built into them a specific skepticism about the role of judges and the social and class interests they serve.

the Convention has been adopted by all member states of the Council of Europe, which includes all 27 members of the European Union.

Article 1, Protocol 1 of the Convention, adopted subsequently in March 1952, addresses the issue of property and its use. It states:

> Every natural or legal person is entitled to the peaceful enjoyment of his possessions. No one shall be deprived of his possessions except in the public interest and subject to the conditions provided for by law and by the general principles of international law.
>
> The preceding provisions shall not, however, in any way impair the right of a State to enforce such laws as it deems necessary to control the use of property in according with the general interest or to secure the payment of taxes or other contributions or penalties.

Several recent analyses have been published which analyse Article 1, Protocol 1 from the perspective of the case law of the European Court of Human Rights (for example Ploeger and Groetelaers 2007; Allen 2007, 2010).[14] Both Allen (2010) and Ploeger and Groetelaers (2007) note that there has been a sharp increase in the number of cases brought before the Court since the 2000, and that '(T)he right to property is one of the most frequently litigated Convention rights' (Allen 2010: 1078).

Ploeger and Groetelaers note that

> [i]n general, the Court leaves the national authorities a wide 'margin of appreciation' to implement social and economic policies. Therefore, it will respect the judgment of the national legislature that the measure is in the general interest, unless that is manifestly without reasonable foundation. (2007: 1431)

At the same time, the Court has developed a trans-European concept of property that is not solely dependent on national laws or definitions; it has 'an autonomous meaning' (Ploeger and Groetelaers 2007: 1427).

What concerns Allen (2010: 1055) is that in its original construction Article 1, Protocol 1 'represented a compromise between liberal and social democrat visions of property … [h]owever, the Court has not been entirely comfortable with that compromise, and it has steadily moved to a liberal construction' of the clause. What does he mean by this? According to Allen

---

14 There is other, though limited, English language literature on this subject, some of it focused on very particular subjects. Take note of, for example, Coban (2004), Fabri (2002) and Schwelb (1964), as well as two guides prepared for jurists by The Council of Europe – Carss-Frisk (2001) and Grgić et al. (2007). For the purposes of this chapter, I rely heavily on to the analysis in Ploeger and Groetelaers (2007) and Allen (2010).

> [t]he liberal view of human rights and democracy concentrates on the protection of individual liberty from the State. It regards expropriation and regulation as interferences in a natural or normal state of affairs … as a general principle, liberal governments would leave decisions regarding the use and disposition of property to the owners … this view … sees the relationship between owners of private property and the State as one of opposition. (2010: 1056)

In contrast,

> [t]he social democratic vision [sees] the … regulation of property in a fundamentally different light … the regulation of property is therefore not a primary threat to human rights: it is the failure to regulate that is the greater threat. For social democrats … regulation *is* the normal state of affairs. (Allen 2010: 1058, emphasis in original)

Ploeger and Groetelaers (2007: 1436) argue that their analysis suggests that 'any involvement with the use of land, so any planning instrument, will constitute an "interference" in the sense of Article 1'. So it is not *if* planning instruments interfere, but rather *how* that interference impacts 'the peaceful enjoyment of the property' and whether it strikes 'a fair balance' of public and private needs and goals in the context of the protections offered by Article 1, Protocol 1.

The notion of 'fair balance' emerges from the Court's decisions. Both Allen (2010) and Ploeger and Groetelaers (2007) focus on a key set of cases that articulate the Court's thoughts on how Article 1, Protocol 1 should be implemented (see also Grgić et al. 2007). Both identify the 1982 case of *Sporrong and Lönnroth v. Sweden* as a watershed shift. For Allen (2010) the case is particularly troubling. As he notes, *Sporrong* 'represents a radical shift in thinking: from reviewing only the lawfulness and the purpose of legislation, the Court would now review the fairness of the legislation' (Allen 2010: 1065). But more importantly, *Sporrong*

> re-aligned the liberal-social democrat balance … The adoption of the fair balance test favoured the liberal model of the right to property … The judgment therefore casts doubt on the social democrat views of property and value in the regulatory State. (Allen 2010: 1066)

As he concludes, Allen emphasizes this point:

> [t]he *Sporrong* judgment therefore subjected regulatory policies to closer human rights scrutiny, especially as it shifted … to the liberal view that an unregulated market is the normal state of affairs, from which all departures require justification. (2010: 1073)

Ploeger and Groetelaers conclude that '[t]he use of land development tools and planning tools will find a boundary in the protection of fundamental right to property' and that 'any involvement with the use of land, so any planning instrument, will constitute an "interference" in the sense of Article 1' (2007: 1436).

Allen (2010: 1067) agrees that the Court extended Article 1, Protocol 1 'to protect property from both direct and indirect interference with rights of property'.[15]

Neo-liberal organizations exist in Europe and advocate a range of policies similar to those advocated by their American counterparts – smaller government, less regulation, lower taxes, flexible labour markets, job creation, strong borders and anti-immigration. But unlike their American counterparts, Europeans seem less focused on land and property rights per se.[16] But still there is significant discourse about property in Europe. According to public officials the mood of landowners is undergoing a seismic shift. Landowners are more willing to challenge local authorities, and in the process a very new relationship is emerging about the balance of the relative rights of the individual and that of society (as embodied by government). And as landowners undertake these challenges, it seems as if the legal system – particularly the decisions of the European Court of Human Rights under the European Convention on Human Rights – is more likely to back them. These changes reflect a spreading sense of a changed European mood, a mood more favourable to markets and property rights and away from the long-standing authority of the central state.[17]

Conclusion

Discourse about property rights is present and strong in both the US and Europe. And in both nations this discourse is shaped by similar liberal (that is, conservative) political ideas.[18] What is different, though, is how the discourse is emerging and the form that it is taking.

---

15 Does this mean that the legal systems and the legal guidelines for spatial planning in the US and Europe are functionally converging? Has *Sporrong* and other key cases litigated before the European Court of Human Rights created a situation similar to that created by the 1922 Penn Coal decision in the US? I believe so. See my detailed discussions about this matter in Jacobs (2008b, 2009).

16 For one exploration of why this might be so, see the discussion in Jacobs 2008b.

17 While I use case studies in France and Norway to make my point, my experience in other parts of Europe (for example Spain), and the breadth of legal cases heard before the European Court of Human Rights, suggest that phenomenon I reference is broader, even if it is not present to the same extent in all western European countries; see Allen (2010) and Ploeger and Groetelaers (2007) for a discussion of these other cases.

18 There is a debate about using the term 'nation' to refer to the European Union, but the term has a both a long history and contemporary usage. For example, since the mid-1800s advocates have urged for the creation of a 'United States of Europe.' Victor Hugo used the phrase in 1849; Herriot, a long-standing member of the French cabinet in the early twentieth century, used the phrase as the title of his 1930 book; Winston Churchill used the term in a speech in 1946, and in 2004 it was the title of a popular book by Reid.

In the US the discourse is very much being pushed by selected advocacy organizations who see property rights as central to their interpretation of the foundational social contract that is the United States. They are the ones who actively work to keep the issue in the media, before legislative bodies, and as a component of formal litigation. In (western) Europe the discourse seems to be present less because of the activism of specific advocacy organizations and is instead a result of changing economic conditions for individual landowners and a mood of economic liberalism they are tapping into which are causing them to rethink what it means to own land and what should be their relationship vis-à-vis government. Some of these landowners are in turn litigating the nature of their rights under Article 1, Protocol 1 of the European Convention on Human Rights and they are finding a sympathetic forum in the European Court of Human Rights.

For many reasons I believe that social debate about property rights – specifically the appropriate balance of individual and social (governmental) rights in land – will continue in the US and Europe, as well as globally.[19]

In the US, it is not clear that state-based property rights laws, either 1990s first generation laws or the more recent anti-*Kelo* laws are having much impact on spatial planning. But this is unlikely to dampen the activism of US property rights advocates. Just as they have, for over two decades, sought to continuously advance their agenda and to learn from their policy experiments, they will, again, learn from their successes and failures with state-based *Kelo* legislation. State-based *Kelo* laws represent the latest, not the final, wave of policy activism on property rights issues by these advocates.

In Europe the situation will be somewhat different. For the most part, advocacy organizations will ignore the types of land and property issues engaged by their US counterparts. But their general levels of activism together with the momentous changes occurring within Europe and relative to Europe's position to the world, will keep a liberal perspective an active part of public and policy discourse. As it does, it is likely to influence broader public and policy discussions and legal decisions about the rights and responsibilities of the individual and the state.

What does all this mean for spatial planning? Spatial planning is premised on the fact that individual decisions over land will not, when cumulated, result in the greater social interest. Therefore there is a need for public action over land. But the discourse over property rights is a powerful one. The idea that there should be strong private property rights minimally subjugated to the state fits with contemporary rhetoric about markets and democracy. To the extent that this discourse prevails it would appear to be difficult to advance spatial planning schemes in the public interest. One challenge for spatial planners is to develop and advance a discourse

See Wikipedia http://en.wikipedia.org/wiki/United_states_of_europe, accessed on 19 June 2011, especially the History sub-section, and the citations noted.

19 See Jacobs (2010) for my most recent general analysis about why this is likely to be true in the US.

about property rights (and spatial planning) that is as compelling as that advanced by liberals. While this may seem difficult, it is essential for planning's future.

Globally, public policy discussion is focused on the economic fiscal crisis, the 'Great Recession', and its systemic impacts at the national and local levels. Countries and communities are having to have unexpected and uncomfortable discussions about the provision of public services as the tax base which supports those services has softened and declined (sometimes significantly and precipitously). It is in this context that it is necessary to project upon the likely future of spatial planning and the way the discourse over property rights will shape it. Ironically I believe it is likely that spatial planning may be reinvigorated as a result of the crisis. Why? Places severely affected by the fiscal crisis will be forced to re-examine any and all tools at their disposal as ways to address oversupply in the housing sector and to facilitate economic and social redevelopment. As communities do this, it is not at all clear what, if any, resistance their citizens will want them to experience, including resistance from their neighbours professing the integrity and strength of their property rights. What is needed now are real solutions to evermore complex problems. And if an assertion of property rights stands in the way of those solutions, it is not clear that it will heeded. So, despite the rhetoric of liberals, spatial planners and others who advocate for the planning process may not need to be defensive about what they do. Spatial planners have the opportunity to mount arguments for planning solutions to contemporary problems. The current dearth of ideas for how to move forward may allow for the serious engagement of these arguments, including arguments about the appropriate role of property rights in modern, democratic, market societies.

References

Adams, J. 1851 (1790). Discourses on Davilia, A series of papers on political history, in *The Works of John Adams*, vol. 6, edited by C.F. Adams. Boston, MA: Little Brown.

Alexander, G.S. 2006. *The Global Debate over Constitutional Property: Lessons for American Takings Jurisprudence*. Chicago: University of Chicago Press.

Allen, T. 2005. *Property and the Human Rights Act 1998*. Oxford: Hart.

Allen, T. 2007. Compensation for property under the European Convention on Human Rights. *Michigan Journal of International Law*, 28(2), 287–335.

Allen, T. 2010. Liberalism, social democracy and the value of property under the European Convention on Human Rights. *International and Comparative Law Quarterly*, 59(4), 1055–1078.

Alterman, R. 1997. The challenge of farmland preservation: Lessons from a six-nation comparison. *Journal of the American Planning Association*, 63(2), 220–243.

Andrew, J.A. 1997. *The Other Side of the Sixties: Young Americans for Freedom and the Rise of Conservative Politics*. New Brunswick, NJ: Rutgers University Press.
Askheim, L.O. and Rødsæther, C. 1998. Norway, in A. Hurndall (ed.) *Property in Europe: Law and Practice*. London: Butterworths, 421–453.
Booth, P. 2003. Promoting radical change: The *Loi Relative à la Solidarité et au Renouvellement Urbains* in France. *European Planning Studies*, 11(8), 949–963.
Booth, P.A. 2005. The nature of difference: Traditions of law and government and their effects on planning in Britain and France, in B. Sanyal (ed.), *Comparative Planning Cultures*. New York: Routledge, 259–283.
Bosselman F.P. and Callies, D. 1971. *The Quiet Revolution in Land Use Control*. Washington, DC: United States Government Printing Office.
Bosselman, F., Callies, D. and Banta, J. 1973. *The Taking Issue: An Analysis of the Constitutional Limits of Land Use Control*. Washington, DC: United States Government Printing Office.
Brands, H.W. 2000. *The First American: The Life and Times of Benjamin Franklin*. New York: Anchor Books.
Brooks, D. 2001. One nation, slightly divisible, *The Atlantic Monthly*, 288(5), 53–65.
Carss-Frisk, M. 2001. *The Right to Property: A Guide to the Implementation of Article 1 of Protocol No. 1 to the European Convention on Human Rights*. Human Rights Handbooks, no. 4. Strasbourg, Belgium: Directorate General on Human Rights and Legal Affairs, Council of Europe.
Çoban, A.R. 2004. *Protection of Property Rights within the European Convention of Human Rights*. Burlington, VT: Ashgate.
Daniels, T. and Bowers, D. 1997. *Holding our Ground: Protecting America's Farms and Farmland*. Covelo, CA: Island Press.
De Soto, H. 2000. *The Mystery of Capital: Why Capitalism Triumphs in the West and Fails Everywhere Else*. New York: Basic Books.
Dubois, M.P. 1998. France, in A. Hurndall (ed.), *Property in Europe: Law and Practice*. London: Butterworths, 179.
The Economist. 2011a (17 March). The Far right in northern Europe: on the march. Populist anti-immigration parties are performing strongly across Northern Europe. [Online]. Available at: www.economist.com [accessed 13 April 2011].
The Economist. 2011b (17 March). What can France's political establishment do about the rise of Marine Le Pen? [Online]. Available at: www.economist.com [accessed 13 April 2011].
Ely, J.W. Jr. 1992. *The Guardian of Every Other Right: A Constitutional History of Property Rights*. New York: Oxford University Press.
Ely, J.W. Jr. 2009. Post-Kelo reform: Is the glass half full or half empty? *Supreme Court Economic Review*, 17(1), 127–150.

Emerson, K. and Wise, C.R. 1997. Statutory approaches to regulatory takings: State property rights legislation issues and implications for public administration. *Public Administration Review*, 57(5), 411–422.

Fabri, H.R. 2002. The approach taken by the European Court of Human Rights to the assessment of compensation for 'regulatory expropriations' of the property of foreign investors. *New York University Environmental Law Journal*, 11(1), 148–173.

Franklin, B. 1907 (1789). Queries and remarks respecting alterations in the constitution of Pennsylvania, in Albert H. Smith (ed.), *The Writings of Benjamin Franklin*, vol. 10. London: Macmillan and Co.

Fukuyama, F. 1989. The end of history? *The National Interest*, 16, 3–18.

Gates, P.W. 1968. *History of Public Land Law Development* (written for the Public Land Law Review Commission). Washington, DC: United States Government Printing Office.

Gottlieb, A.M. (ed.) 1989. *The Wise Use Agenda*. Bellevue, WA: The Free Enterprise Press.

Grgić, A. et al. 2007. *The Right to Property under The European Convention On Human Rights*. Human rights handbooks, no. 10. Strasbourg, Belgium: Directorate General on Human Rights and Legal Affairs, Council of Europe.

Hamilton, A., Madison, J., and Jay, J. 1961. *The Federalist Papers*. New York: Mentor Books. (Original work published 1788).

Herriot, E. 1930. *The United States of Europe*, translated by Reginald J. Dingle. New York: The Viking Press.

Huntington, S.P. 1997. *The Clash of Civilizations and the Remaking of World Order*. New York: Touchstone.

Höbelt, L. 2003. *Defiant Populist: Jörg Haider and the Politics of Austria*. West Lafayette, IN: Purdue University Press.

Ignazi, P. 2003. *Extreme Right Parties in Western Europe*. Oxford: Oxford University Press.

Jackson, K.A. 2011. *2011 Report—International Property Rights Index*. Washington, DC: Property Rights Alliance. [Online]. Available at: www.internationalpropertyrightsindex.org [accessed 22 September 2011].

Jacobs, H.M. 1995. The anti-environmental, 'wise use' movement in America. *Land Use Law and Zoning Digest*, 47(2), 3–8.

Jacobs, H.M. 1998a. The impact of state property rights laws: Those laws and my land. *Land Use Law and Zoning Digest*, 50(3), 3–8.

Jacobs, H.M. 1998b. The 'wisdom', but uncertain future, of the wise use movement, in H.M. Jacobs (ed.), *Who Owns America? Social Conflict over Property Rights*. Madison: University of Wisconsin Press, 29–44.

Jacobs, H.M. 1999. *State Property Rights Laws: The Impacts of Those Laws on My Land*, Policy Focus Report. Cambridge, MA: Lincoln Institute for Land Policy.

Jacobs, H.M. 2003. The politics of property rights at the national level: Signals and trends. *Journal of the American Planning Association*, 69(2), 181–189.

Jacobs, H.M. 2006. *The Taking of Europe: Globalizing the American Ideal of Private Property?* Working Paper. Cambridge, MA: Lincoln Institute of Land Policy.
Jacobs, H.M. 2008a. L'engrenage de la croissance urbaine: La place de la propriété dans la planification urbaine. *Etudes Foncières*, 132, 12–16.
Jacobs, H.M. 2008b. The future of the regulatory taking issue in the U.S. and Europe: Divergence or convergence? *Urban Lawyer*, 40(1), 51–72.
Jacobs, H.M. 2009. An alternative perspective on U.S. – European property rights and land use planning: Differences without any substance. *Planning and Environmental Law*, 61(3), 3–12.
Jacobs, H.M. 2010. Social conflict over property rights: The end, a new beginning or a continuing conversation? *Housing Policy Debate*, 20(3), 329–349.
Jacobs, H.M. and Bassett, E.M. 2010. After *Kelo*: Political rhetoric and policy responses. *Land Lines*, 22(2), 14–20.
Jacobs, H.M. and Bassett, E.M. 2011. All sound, no fury? The impacts of state-based *Kelo* laws. *Planning and Environmental Law*, 63(2), 3–8.
Kayden, J.S. 2002. Tahoe-Sierra Preservation Council v. Tahoe Regional Planning Agency: About more than moratoria. *Land Use Law and Zoning Digest*, 54(10), 3–5.
Kayden, J.S. 2004. Charting the constitutional course of private property: Learning from the 20th century, in H.M. Jacobs (ed.), *Private Property in the 21st Century*. Northampton, MA: Edward Elgar, 31–49.
López, E.J., Jewell, R.T. and Campbell, N.D. 2009. Pass a law, any law, fast! State legislative responses to the Kelo backlash. *Review of Law and Economics*, 5(1), 101–135.
Marzulla, N.G. 1996. Property rights movement: How it began and where it is headed, in P.D. Brick and R. McGreggor Cawley (eds), *A Wolf in the Garden: The Land Rights Movement and the New Environmental Debate*. Lanham, MD: Rowman and Littlefield, 39–58.
Mayer, N. 2003. Le Pen's comeback: The 2002 French presidential election. *International Journal of Urban and Regional Research*, 27 (2), 455–459.
Mead, W.R. 2011. The tea party and American foreign policy. *Foreign Affairs*, 90(2), 28–44
Miniter, R. 1994. You just can't take it anymore: America's property rights revolt. *Policy Review*, 70, 40–46.
Montopoli, B. 2010. Tea party supporters: Who they are and what they believe. [Online]. Available at: www.cbsnews.com [accessed 9 April 2011].
Morris, A.P. 2009. Symbol or substance? An empirical assessment of state responses to *Kelo*. *Supreme Court Economic Review*, 17, 237–278.
Moss, E. ed. 1977. *Land Use Controls in the United States: A Handbook on the Legal Rights of Citizens by the Natural Resources Defense Council, Inc.* New York: The Dial Press/James Wade.
Nadler, J., Diamond, S.S. and Patton, M.M. 2008. Government takings of private property: *Kelo* and the perfect storm, in N. Persily, J. Citrin, and P. Egan (eds),

Public Opinion and Constitutional Controversy. Oxford University Press, 287–310.

Newman, P. and Thornley, A. 1996. *Urban Planning in Europe: International Competition, National Systems, and Planning Projects*. London: Routledge.

Ploeger, H.D. and Groetelaers, D.A. 2007. The importance of the fundamental right to property for the practice of planning: An introduction to the case law of the European Court of Human Rights on Article 1, Protocol 1. *European Planning Studies*, 15(10), 1423–1438.

Reid, T.R. 2004. *The United States of Europe: The New Superpower and the End of American Supremacy*. New York: Penguin

Revell, K.D. 1999. The road to Euclid v. Ambler: City planning, state-building, and the changing scope of the police power. *Studies in American Political Development*, 13(1), 50–145.

Roddewig, R.J. and Duerksen, C.J. 1989. *Responding to the Takings Challenge: A Guide for Officials and Planners*. Planning Advisory Service Report No. 416. Chicago, IL: American Planning Association.

Sagalyn, L.B. 2008. Positioning politics: *Kelo*, eminent domain, and the press, in Richard Francis Keevey and Timothy N Castano (eds), *Land and Power: The Impact of Eminent Domain in Urban Communities*. Princeton, NJ: Policy Research Institute for the Region, Woodrow Wilson School of Public and International Affairs, Princeton University, 49–74.

Salkin, P.E. 2001. *Trends in Land Use Law from A to Z: Adult Uses to Zoning*. Chicago, IL: American Bar Association.

Schain, M. et al. 2002. *Shadows over Europe: The Development and Impact of the Extreme Right in Western Europe*. New York: Palgrave.

Schlatter, R. 1951. *Private Property: The History of an Idea*. New Brunswick, NJ: Rutgers University Press.

Schwelb, E. 1964. The protection of the right to property of nationals under the First Protocol to the European Convention on Human Rights. *American Journal of Comparative Law*, 13(4), 518–541.

Shorto, R. 2011. Marine Le Pen, France's (kinder, gentler) extremist, *The New York Times Magazine* (01 May). [Online]. Available at: www.nytimes.com [accessed 1 May 2011].

Somin, I. 2009. The limits of backlash: Assessing the political response to Kelo. *Minnesota Law Review*, 93(6), 2100–2178.

Teles, S.M. 2008. *The Rise of the Conservative Legal Movement: The Battle for Control of the Law*. Princeton, NJ: Princeton University Press.

Turner, F.J. 1893. The significance of the frontier in American history, *Report of the American Historical Association*. 199–227.

Wikipedia 2011a. Swedish General Election, 2010. [Online]. Available at: http://en.wikipedia.org/wiki/Swedish_general_election,_2010 [accessed 13 April 2011].

Wikipedia. 2011b. Tea Party Movement. [Online]. Available at: http://en.wikipedia.org/wiki/Tea_Party_movement [accessed 9 April 2011].

Wolf, M.A. 2008. *The Zoning of America: Euclid v. Ambler*. Lawrence, KS: University of Kansas Press.

Woods, D. 2009. Pockets of resistance to globalization: The case of the Lega Nord. *Patterns of Prejudice*, 43(2), 161–177.

Case interviewees

France

Bourdat, E. 2006 (01 February). Chargée de Mission, Mission Interministérielle d'Aménagement du Littoral, Préfecture de région du Languedoc-Roussillon.

Gambier, J-P. 2006 (31 January). Chef du Service Foncier, Montpellier Agglomération.

Girault, I. 2006 (03 February). Directeur, Agence d'Urbanisme et de Développement de la Région Nîmoise.

Norway

Aasen, I. 2006 (30 May). Senior Adviser in the Department of Forest and Natural Resource Policy, Ministry of Agriculture and Food (Landbruks- og matdepartementet).

Falleth, E. 2006 (30 May). Member of the Environmental Planning Research Group of the Norwegian Institute for Urban and Regional Research (NIBR).

Hjelle, A.M.A. 2006 (31 May). Assistant General Secretary of the Norwegian Mountain Touring Organisation (Den Norske Turistforeningen).

Jensen, Dr. R.H. 2006 (01 June). Private planning consultant and former academician.

Rasmussen, K.F. 2006 (01 June). Senior Adviser in the Housing and Building Department, Ministry of Local Government and Regional Development (Kommunal- og regionaldepartementet).

Sevetdal, H. 2006 (02 June). Professor in the Department of Landscape Architecture and Spatial Planning (ILP), Norwegian University of Life Sciences (UMB).

Tesli, A. 2006 (30 May). Research Director of the Environment and Development Group of the Norwegian Institute for Urban and Regional Research (NIBR).

Chapter 5
Managing Riverside Property: Spatial Water Management in Germany from a Dutch Perspective

Thomas Hartmann and Tejo Spit

Why is it so difficult for water managers and spatial planners to collaborate in managing riverside properties? In this chapter, we investigate this question from a legal perspective. Owing to recent trends in European environmental policy – in particular with respect to flood risk management – water managers need to influence land use behind the dykes. Thus, they are encroaching on the traditional arena of spatial planners. Water management and spatial planning have essentially different approaches to managing land. In this contribution, we describe the different modes of governance of both institutions and explain how these are institutionally embedded in the law. We provide from an outsider's, Dutch, perspective new insights into the entrenched particularities of German legislation: namely the role of condition-based and of performance-based legislation. The underlying research interest concerns the way in which particular types of legislation influence modes of governance of spatial planning and water management.

European and German Flood Risk Policy and Legislation

While we write this article in early 2011, high water waves are discharging into the Rivers Rhine, Oder, Danube, Elbe, and Maas. Almost every year, a major flood occurs somewhere in Europe. In addition, Australia recently endured a major flood and flash floods caused damage in Canada, Brazil, and Sri Lanka. As a consequence, water managers experience a shift in the approach to coping with flood events. The flood events in 1993 and 1995 on the River Rhine initiated a paradigm shift from infrastructural flood protection to more spatial flood risk management (Moss and Monstadt 2008: 63; Mostert and Junier 2009: 4961; Roth and Warner 2007). The Netherlands and Germany have played a key role in that paradigm shift (LAWA 1995: 8; UBA 1999: 25–27). '"Space" now became the keyword in Dutch flood policy… While earlier measures were restricted to the areas between dykes, this time Rijkswaterstaat [the Dutch water management authority] was going to intervene behind dykes' (Roth and Warner 2007: 520–521). Water should no longer be excluded; it should rather be 'accommodated' (Wesselink

2007: 243). This idea did not fit in with their working paradigm. Previously, the idea of managing water with technical and infrastructural-engineering solutions dominated the water management thinking. Water management has not been prepared for a catchment-wide planning of flood risk management (Ison et al. 2007: 500). The dykes have been regarded as the boundaries between the water and water management. But now, a dyke is no longer the boundary or the line of defence in the 'battle against the water' (Wiering and Immink 2006: 423) or a line of demarcation, where the dominance of water management ends and the playground of spatial planning begins.

The way the law is phrased not only determines the legal consequences of certain actions, but also has a crucial impact on the working paradigms – the modes of governance – of the executive administration. Thus, as we assert spatial planners and water managers need to know the impact of different types of legislation on their work, and on their mutual cooperation. The modes of governance, we argue, are entrenched in the way the law is phrased. This state of affairs becomes obvious in managing riverside properties in spatial water management.

European Water Policy

For the first time, in the year 2000 the European Water Framework Directive demanded a spatial water management. In order to achieve a 'good [ecological] status' of water bodies, whole river catchments should be attended to – instead of just the bodies of water and the embankments (2000/60/EC). This directive did not explicitly include flood protection, but in 2007 the European Union released the European Flood Risk Management Directive (Mostert and Junier 2009: 4963). Accordingly, water managers now need to prepare flood risk management plans up to 2015 (article 7 I EU-FD). These affect spatial planning in riparian landscapes (article 7 III EU-FD; Mostert and Junier 2009: 4966) and restrict urban developments in these areas (Albrecht 2007: 501). Future water managers need to plan behind the dykes, not just between them (here the term 'between the dykes' is used instead of 'in the floodplains' to highlight the difference from 'behind the dykes'). Now they have to plan in the field of spatial planning, which in turn has to incorporate the new role of water management. For a long time, the relationship between spatial planning and water management has been conflicting, but well-rehearsed (Moss 2009: 54; Wiering and Immink 2006: 424). Recent European water policy intensifies the friction between spatial planning and water management.

Particularly in Germany, water engineers struggle with managing riverside properties behind the dykes. In those areas, spatial planners develop binding land-use plans in a countervailing and nested hierarchical system of plans, based on balancing different spatial issues. Each of these needs to pass through 'the needle eye of balancing spatial planning decisions' (Durner and Ludwig 2008: 467). German water managers struggle because they are not used to this way of planning (Durner and Ludwig 2008: 461). In countries with a strong institutional division between the water and planning sectors – such as in Germany – integrative water management is difficult to achieve (Moss 2004: 86). 'Water authorities accustomed to operating with regulatory instruments along vertical paths of the administrative hierarchy will in the future need to cultivate more intensively forms of horizontal cooperation' (Moss 2004: 90). This horizontal cooperation requires some rethinking. Different social constructions of water need to be taken into account. In practice, water managers tend not to understand that water is more than a biophysical entity, as it is has been regarded historically for a long time (Ison et al. 2007: 503). A claim like 'water and its physical and social characteristics create interdependencies that must be taken into account by humans who then conceptualize particular ways of understanding water' (Ison et al. 2007: 503) is simply incomprehensible from a technocratic view. In this respect, the water sector tends to be inflexible and not prepared to adapt to recent changes in environmental, economic or social circumstances (Pahl-Wostl 2002 quoted in: Ison et al. 2007: 501). Wesselink describes the situation of the flood defence system as locked-in politically, where 'it is unable to escape its control paradigm for flood risks' (Wesselink 2007: 246). Water managers perceive a gap; on the one hand particular European directives, but also national attempts of policymakers and the scientific community, claim that water managers need to plan in an integrative manner and implement their measures and plans catchment-wide. On the other hand, the federal German water law (*WHG* = '*Wasserhaushaltsgesetz*') does not provide essentially new instruments for water managers to do so (Knopp 2010: 124).

The institutional setting of water management in Germany sustains a traditional control paradigm in which particular characteristics of the German water legislation are entrenched. German water managers are unable to escape their control paradigm. Germany has already struggled with integrating the regulations resulting from the Water Framework Directive in spatial planning (Moss 2009: 57), but the European Flood Risk Management Directive seems to cause even more difficulties, particularly because it includes restrictions and even prohibitions for building in the floodplains (Albrecht and Wendler 2009: 611). How can traditional water management be better prepared for the new spatial water management? In the next few years, this question will become crucial (and not only from a legal perspective); but the law prescribes that flood risk management plans have to be prepared until 2015 (§ 75 WHG).

By looking from an outside perspective – from the Netherlands – we can identify a fundamental legal constraint on German water management that hinders planning behind the dykes, and that explains an entrenched paradigm of controlling property in land. We therefore elaborate the specifics of two different types of legislation in the water and planning sectors: condition-based and performance-based legislation. This distinction clarifies the deep relationship between the law and modes of governance of administrations.

Condition-based and Performance-based Legislation

At first glance, one reason for the struggle of German water managers in implementing the Flood Risk Management Directive can be seen in German federalism and its reform in 2006. Previously, the national federation in Germany had only had a confined legislative competence in the water sector; the *Länder* had their own water laws. Consequently the implementation of the European Water Framework Directive in national law was very complex and slow in Germany before the reform of federalism (Steindorf and Häberle 2010: no. 13a). This reform from 2006 entitled the federation to regulate the water sector fully. Thus, the government passed a new Federal Water Act. This act, however, has been criticized because it was not detailed enough (it does not have the regulatory depth of a '*Vollregelung*' – a comprehensive regulation); the federal states would still need to pass additional state-level legislation to make the regulations concrete and to implement them (Pape 2010: no. 64). From this perspective, some confusion of German water managers implementing the European flood risk policy seems understandable. But this is just the first glance. A closer view reveals that the new Federal Water Act marks a shift in the type of legislation: from condition-based to performance-based. It is not that the new legislation is too vague, but rather that it is a different type of legislation.

Performance-based and condition-based are two different types of legislation. The term *performance-based* is derived from the US-American discussion of the topic (see for example Coglianese et al. 2002); in the German discussion, the word '*finale Gesetzgebung*' is used – but there are hardly any suitable English-language translations. Thus, here the term *performance-based legislation* is used, since it covers what is meant by '*finale Gesetzgebung*'. The term *condition-based* is used to describe the other type of legislation. The term is derived from the German word '*konditionale Gesetzgebung*' (see Fonk 2010) – but the English word *conditional* does not match the German term precisely. Lacking a better alternative, condition-based legislation is the term we have used for what is meant by the German concept. To clarify these terms, their characteristics and differences between them are explained below.

This distinction between performance-based and condition-based legislation is not between Anglo-American case law (common law) and the Romano-Germanic codified law of continental Europe (Tetley 1999: 592). The distinction is rather within codified law. This aims to regulate cases beforehand, abstractly, comprehensively, and exclusively (Fonk 2010: 626). Performance-based legislation and condition-based legislation differ in the way in which a regulation governs: either by describing precisely what actions the addressed entities have to take, under what conditions, and in what situations, or by incorporating 'the regulation's aim into the language of the rule' (Coglianese et al. 2002: 706). The latter – performance-based legislation – focuses on the result of an executive activity; the purpose or aim of a regulation is determined and it is left to the administrator how those aims are to be met. Condition-based legislation emphasizes procedures of administration; the aim of a regulation is not determined, but the conditions under which a certain legal consequence is to be applied by the administration (Fonk 2010: 629).

Performance-based legislation (‘*finale Gesetzgebung*’) implies that the planning process is oriented towards set planning aims. All the instruments and procedures the law provides and determines are designed to identify and alleviate certain deficits in a determined period of time (Albrecht and Wendler 2009: 609). Performance-based legislation sets a frame, namely the aim, but leaves the public administration considerable scope for decisions (Durner and Ludwig 2008). The judiciary control within performance-based legislation is confined to the balancing processes carried out by the administration (Albrecht 2007: 98). This balancing requires well-elaborated ‘rational’ decisions that are consistent within the system of programs and plans (Breuer 2007: 505).

Condition-based legislation (‘*konditionale Gesetzgebung*’) is the contrasting model of legislation: regulations are formulated in if-then relationships. This formulation is usual in civil and penal law – if one drives a car X kilometres per hour too fast, one has to pay Y. The facts of a case are usually comprehensively regulated and are subject to judicial control (Albrecht 2007: 98; Breuer 2007: 505). Consequently the administration has very limited scope in a decision. Condition-based legislation determines precisely ‘what happens when’. It is not necessary to name the purpose of the regulation in the text of the law (although there is a trend also to do this) (Fonk 2010: 629).

As a consequence, the two types of legislation create different types of legal certainty: performance-based regulations need to be benchmarked according to the set aim, whereas condition-based regulations need to be assessed according to the legally-defined conditions set out in the law. Condition-based legislation requires a rather precise administration that adheres to the if-then conditions, albeit with discretion. Thus, a condition-based administrative regime is used to act within the system of the law and expects the law to determine in detail what to do, and when. In the case of conflict, such an administration would look to the law to resolve it. A performance-based administrative regime is used in a more self-determined way. In the case of conflict, the administration has to provide a sound and rational line of argument for its actions in the process. Such an administration acts more politically. Thus performance-based legislation requires a more flexible administration, willing to balance issues to achieve consensus and compromise to achieve the aim. As a result, the type of legislation has a crucial impact on the mode of governance (Fonk 2010: 631). How does the German situation of water management and spatial planning relate to the two types of legislation?

German Planning and Water Law

German water right is traditionally based on condition-based legislation (Breuer 2007: 505, Albrecht 2007: 98), whereas German planning law pursues the model of performance-based legislation (Albrecht 2007: 98). It has to be emphasized that the assignment of water sector and planning to condition-based and performance-based legislation is not a black-and-white distinction; there is a rather grey zone, where elements of both types of legislation can be found in a law.

The German Federal Water Act – particularly before the last reform in 2010 – predominantly contained regulatory instruments for controlling the use of water and for the protection of and against water for particular cases (Knopp 2010: 49). The regulations in German water laws are quite detailed and leave little scope for variation at the executive level. For example the granting of permission and allowance of water use ('*Erlaubnis und Bewilligung*') are typical condition-based regulations, because they depend predominantly on procedural 'if-then' regulations. In the Federal Water Act, the concept of balancing interests is not so dominant (although it is becoming more important, also for water legislation) (see for example Knopp 2010: 54 about § 6 WHG, the basic principles of water management). Particular regulations regarding urban developments in the floodplains determine precisely the conditions under which a building permit may be granted or must be denied. Article 78 of the German water law contains a detailed list of land uses that may be permitted in formally-registered inundation zones; nine detailed criteria for zoning plans within such areas are listed; finally, the conditions that exceptional building in floodplains needs to meet are set out. Although German water law is not purely condition-based most regulations are, so that this law can be assigned to this type of legislation.

German planning law has a strong tendency towards performance-based legislation. This tendency can be seen for example in article 1 of the German Building Code (the relevant law for municipal land-use planning), where the (performance-based) aims of municipal land-use planning are determined (see also Fonk 2010: 628): Land-use plans have to achieve a sustainable urban development, including social, economic, and environmental aspects as well as intergenerational justice: the land use should be socially fair, and the plans should care for a human environment and protect the natural resources; flood protection is also mentioned as an issue that needs to be considered in the balancing process (§ 1 VI no. 12 BauGB) (*BauGB = Federal German Building Code*). Also the German Regional Planning Act (*ROG = 'Raumordnungsgesetz'*) – also relevant for German spatial planning – determines content-oriented principles for spatial development (§ 2 ROG). German planning law is even seen as the prototype of performance-based legislation in Germany (Albrecht 2007; Fonk 2010). So, spatial planning and water management adhere to different types of legislation. Thus, following the arguments above, different modes of managing land can be expected.

Managing Riverside Property

The provision of flood protection has always been the responsibility of water management (Reinhardt 2008: 472). Water managers used to derive a predominant position from this responsibility. Between the dykes, this arrangement works well.

The responsibilities and instruments of water managers for the area between the dykes grew during the revisions and reforms of the German water laws in the last decade. From 2005, the '*Hochwasserschutzgesetz*' (the Federal Flood Control Act) has strengthened the power of water managers to restrict urban development in these areas. The recent reform from March 2010 even enables water managers to forbid further urban development in inundation zones (which was not previously the case, see also Hartmann 2009). But, behind the dykes, the formal competences of water management are still vague (Moss 2004: 85). Although the German legislator made an attempt to strengthen the position of water managers behind the dykes by introducing '*überschwemmungsgefährdete Gebiete*' (potentially floodable areas) (article 31c WHG 2005), the category was abandoned in 2010, because it was thought to be too weak (Reinhardt 2008. 473). The new Federal Water Act of 2010 provides no hard instruments for water managers to plan behind the dykes. Now, European flood risk policy demands water management that governs space behind the dykes (Moss 2009: 56). How do water managers usually approach land-use regulation?

Modes of Property Governance

'*Optimal wäre, das Land ist Eigentümer von Retentionsflächen*' – it would be best if the state became the proprietor of retention sites (State water manager in Saxony-Anhalt in an interview 2008). Water managers usually aim at full ownership of riverside properties. Technical needs and essential requirements for land uses in retention areas and polders justify this position (Schätzl and Hoffmann 2006: 10), but it causes conflicts with landowners (Reinhardt 2003), who do not want to sell their land (for example OVG Koblenz 2004). Land acquisition often protracts the realization of a project crucially, as several studies and projects conducted with students of the School of Spatial Planning at TU Dortmund University have confirmed (Project F01 2006/2007, Project F01 2008/2009). To acquire the required land, expropriation is almost never used in practice – although this instrument is important for water managers as a hypothetical possibility; prevalent methods are freehand purchase, other trade-offs and exchanges of land for requested sites, and, most commonly, rural-land readjustment (UBA 2003: 121). Most of these methods are expensive, time consuming and administratively complex. Sometimes rural-land readjustments can even last decades (Project F01 2008/2009). A water manager in Saxony-Anhalt affirmed: '*das Land müsste die Flächenausweisen, kaufen und dann eventuel lan Landwirte verpachten*' – the state has to designate, buy, and lease the land back to farmers (municipal water manager in Saxony-Anhalt in an interview in 2008). But the implementation of a flood protection measure on land currently used as farmland depends very much on the consent of the farmers who own the land: not only is there a constitutionally guaranteed protection of private property, but also farmers are usually well organized and form a politically-powerful lobby (Gerdes 2008: 103). Typically, farmers oppose selling their land for retention measures. But there are alternatives: Farmers along the River Lippe in North Rhine-Westphalia would be willing to allow temporary inundation of their farmland if they retained the ownership (Project F01 2008/2009). In this vein, a retention area in rural Rhineland-Palatinate has been realized based on an easement (Project F01 2006/2007). However, in general water managers pursue the full ownership or at least extensive control of the land.

Now, the law demands that water managers plan in the hinterland of the dykes, but the law does not provide instruments to enforce water management plans there. Water managers struggle with this situation. Indeed, the Federal Building Code prescribes that spatial planning recognize aspects of flood protection in their plans, but since the state is not in general liable for flood damage (Reinhardt 2004), and since spatial planning also has to balance other issues – economic development, social fair-land policy, environmental issues, and so forth (article 1 V BauGB) – conflicts between planners and water managers can be expected if planners give other sectoral planning aspects a higher priority in certain cases. An official at the agency for regional planning at the state level of Saxony-Anhalt confirmed that flood protection is '*ein wichtiges Thema in Landesentwicklung. Es ist aber nicht wichtiger als Andere*' – an important issue in regional planning at the state level, but it is not in general more important than other topics (interview in 2008). Water managers feel powerless behind the dykes, because they cannot prevent further urban development there (Moss 2004: 85–87). How can water managers approach this new situation best?

Planning behind the dykes is the profession of spatial planners, encompassing regional planning and municipal land-use planning (zoning). Spatial planning should coordinate different sectoral plans and different land-use demands. Spatial planning claims to be the comprehensive and meta-disciplinary institution to resolve spatial conflicts (Söfker 2009: X–XI). Instead of controlling the land by aiming at its ownership, planners influence the activities on private property by preparing plans that are binding on authorities ('*behördenverbindlich*') before finally imposing restrictions for private property in binding zoning plans. Already the basic interests of spatial planning are different from water management. Ideally, spatial planning does not have its own special spatial interest (Krautzberger 2007: marginal 28), but balances the different spatial issues emerging from different sectoral plans (§ 1 VII BauGB and § 1 I ROG). Whereas sectoral planning is expected to formulate specific spatial demands for the particular sector, land-use plans have to achieve multiple aims simultaneously: a sustainable urban development, including social, economic, and environmental aspects as well as intergenerational justice; all public and private interests must be balanced fairly among and against each other (§ 1 BauGB). If two or more issues conflict incompatibly with each other, planners must 'balance issues away' ('*Belange wegwägen*') or shift to other administrative procedures such as determining special requirements for building permits (Paul and Pfeil 2006: 510, Stüer 2005: 3454). Local self-government and the principle of subsidiarity is the legal basis for this capacity to balance spatial interests among and against each other (Stüer 2005: 2991). In cases of conflict, sectoral planners may insist on their supralocal sectoral competences (Stüer 2005:

2990), but even then, spatial interests of municipalities have to be considered in the regional sectoral-planning decision (Stüer 2005: 2995, Paul and Pfeil 2006: 506). Finally, no side is entitled to decide exclusively; everyone decides to some extent (Stüer 2005: 2991).

Institutional Settings

So, there is a tension between sectoral planning and spatial planning owing to the strong sectoral division of responsibilities. This is a 'familiar feature of public administration in Germany' (Moss 2004: 85) as well as in the Netherlands (Spit and Zoete 2009: 20–21). The German planning system draws an essential institutional distinction between spatial planning and water management (Moss 2009: 59). Spatial planning is comprehensive and meta-disciplinary; water management is specific and sectoral (Stüer 2004: 416; Moss 2009: 60). These different modes of governance are expressed in different languages and ways of thinking and working; in the past they have already burdened the collaboration of spatial planners and water managers (Moss 2009: 58). Timothy Moss describes the steering approach of spatial planning as relying on balancing and bargaining between different spatial demands, and of implementing aims with the help of other stakeholders, and the approach of water management as top-down (Moss 2009: 60). But the traditional patterns of land management of water managers no longer work properly behind the dykes: 'Spatial planning landed the [water management] department in an unknown arena where project design does not take place in a top-down fashion, but many interests are involved in a consensual process of negotiation' (Roth and Warner 2007: 520–521). In addition, water managers have a good financial budget, in contrast with spatial planners. This capacity becomes obvious when one enters the main entrance of a water management agency. One will often be immediately overwhelmed, because many water management agencies reside in beautiful historical buildings, with large entrance halls and wide staircases. In contrast, if one enters the section for spatial planning in a municipal authority, one might feel some compassion for the employers. Planning agencies are often located in functional buildings, built in the 70s or 80s, sometimes renovated because of PCB or asbestos. The difference in the financial means of water managers and spatial planners also accounts for the different steering approaches. Spatial planners try to regulate space with a relatively moderate budget.

The relationship between water management and spatial planning has been widely discussed in the Netherlands in terms of facet planning versus sector planning (see for example Spit and Zoete 2009: 19–21). Whereas water management – as sector planning – aims to control and regulate its sector, spatial planning – as facet planning – aims at the coordination and integration of different sector activities. The friction between spatial planning and water management emerges from the fact that sector planning usually has executive power and money; whereas facet planning wants to coordinate and integrate the different sectoral plans in (comprehensive) facet planning by such 'soft' means as mediation, cooperation, communication, and negotiation. Thus, sectoral and facet planning differ essentially in their mode of governance. However, the distinction between them is not black-and-white with sharp boundaries. Rather, sectoral and facet planning are important analytical categories that have in practice become less important, because nowadays traditional sector planning institutions also plan like facet planners (Spit and Zoete 2009: 20).

The Dutch Perspective

In Germany, the discussion about the relationship between spatial planning (as facet planning) and sectoral planning is still lively. In particular, in the water sector the question of '*wer regiert hier eigentlich wen*' – 'who governs whom' (Stüer 2004: 420), emerges each time a new regulation is introduced in water or planning law. New regulations are discussed in terms of a clear definition of competencies between sector and facet planning rather than merging competences. Thus, the analytical distinction in sector planning and facet planning is important in Germany also. In the Netherlands, the relationship between spatial planning and sectoral planning has become much more complex in the last few decades (Spit and Zoete 2009: 20), since the Dutch government merged the ministry responsible for spatial planning with the water ministry.

It is notable that the Dutch legal tradition is not as strongly condition-based as in Germany. Fonk asserts that the strong German orientation towards condition-based legislation is relatively unique in Europe – Germany 'ploughs a lonely furrow' ('*steht allein auf weiter Flur*') (Fonk 2010: 629). The discussion about condition-based and performance-based legislation is not as lively in the Netherlands. There, the discussion about water management behind the dykes seems to be conducted more pragmatically in terms of the coordination between environmental policy and spatial planning (see for example Glasbergen and Driessen 2002). The German discussion, in contrast, is more dogmatic with respect to the two types of legislation. There are some indicators that water laws in the Netherlands are more dominated by performance-based legislation; although Dutch planning law incorporates both condition-based and performance-based regulations, the Dutch government announces on its website, that the Spatial Planning Act (*WRO = wet op ruimtelijke ordening*) '*doelen vooraf stelt*' (Rijksoverheid 2011) prescribes aims. It seems that Dutch planning law – the (WRO) – is developing increasingly into performance-based legislation. According to Ang et al., in an article on the Dutch performance-based approach to building regulations and public procurement, the Netherlands even had a major influence on the type of legislation at the European Union level in 1985 (Ang et al. 2005: 108). However, there is not much literature on the Dutch legal traditions regarding the topic of performance-based and condition-based legislation. Finally, condition-based legislation seems to be more firmly entrenched in Germany than in the Netherlands.

Types of Legislation and Modes of Governance

The analysis of performance-based and condition-based legislation provides an explanation of why this lock-in is so entrenched. But how has it developed in this way?

Suitable Types of Legislation

German water law is one of the oldest German legislations for environmental goods (Nisipeanu 2008: 87). Early water laws regulated matters of all kinds with respect to water use and water management. Originally, water management served predominantly economic purposes: safeguarding shipping, regulating discharge, providing flood protection to make floodplains economically usable or maintaining and increasing the value of river-adjacent land (Nisipeanu 2008: 88). Users of riparian land have been dependent on water management. Particularly in the nineteenth century, water managers became used to resolving problems of all kinds with technical solutions (Nisipeanu 2008: 88–89): examples are the large and famous river reconstruction on the Oberrhein by Johann Gottfried von Tulla, but also the reconstruction of the River Emscher to provide a sewage system for the industrializing Ruhr area. In Germany, water law institutionalized this technocratic approach and supported it by assigning water issues to governmental institutions in the early nineteenth century. Whereas at the beginning, the various water laws regulated the different private uses of water and regarded the maintenance of water bodies as a civil duty, in time the latter became a task of the public administration. Because of the importance for trade and the military, the *Reich* became the proprietor of the large waterways (Nisipeanu 2008: 89). This reflection on the emergence of water legislation in Germany illustrates how the hierarchical approach of water management has grown in the course of time. Water legislation in Germany has not had the same inherent urgency as in the Netherlands. There, water management has always been an essential issue of providing living conditions for the people (Needham 2007: 27), which led to a very strong and long tradition of water management at the local scale; in Germany, procedural aspects and territorial control were more important.

German planning legislation, in contrast with water legislation law, is still rather young. Regulations about the use of land were rare until the industrialization around the middle of the nineteenth century (Söfker 2009: X). The aim of planning law was to alleviate urban disorder and provide planned future urban development. Regional planning developed out of the urban planning law and evolved as its own discipline. The German Building Code, however, emerged from urgency, namely as an instrument to bring order in the urban development after the Second World War (Söfker 2009: X). So, both types of legislation – condition-based and performance-based – have their justification. Depending on the purpose of the regulation, one is sometimes better suited than the other. Performance-based legislation is more flexible, it seems to be better suited for emergencies; whereas condition-based legislation is suited to stabilize institutional power relations.

Towards Performance-based Water Legislation

There is a trend in European water legislation – actually all European legislation on environmental issues – towards performance-based legislation (Durner and Ludwig 2008: 458; Breuer 2007: 505). The Water Framework Directive (2004) and the European Flood Risk Management Directive (2007) determine particular aims for managing water (Albrecht and Wendler 2009: 609) – the 'good status' of water quality (article 4 EU-WFD) and the 'tolerable flood risk' (article 1 EU-FD). Glasbergen and Driessen explain this trend by the paradigm shift in environmental policy in general from technically-oriented government towards integrative governance (Glasbergen and Driessen 2002: 3–4). Another explanation could simply be the competences of the European Union. EU directives need to be implemented via the national laws of the member states. Thus, it is more appropriate to determine the aim of a regulation than use condition-based regulations.

In 2010 Germany incorporated the EU Flood Risk Management Directive almost word-for-word into German water legislation. The flood risk management plan (§ 75 WHG) replaced the former flood protection plans ('*Hochwasserschutzpläne*') (§ 32 d WHG, version 2005). Since the flood risk management plan adheres to performance-based legislation, and the flood protection plan was more of a condition-based regulation, the new instrument determines fewer specifications and focuses more on the purpose of the plan than on the specific conditions to be met by the instrument (Reinhardt and Czychowski 2010: § 75 WHG, No. 11). This is probably the best illustration of the shift in German water legislation. But the German Federal Water Act includes both regulations according to the new paradigm of performance-based legislation, and instruments according to condition-based legislation: and these two kinds of instruments have not yet been integrated in a consistent – either condition-based or performance-based – system. This situation can be illustrated with the example of formal inundation zones: ('*Überschwemmungsgebiete*'). These are the water managers' most important instruments for confining urban development near water. They are in this respect an instrument to influence spatial planning. The inundation zones formally declare restrictions for land uses on riverside properties. Since 1996, when it was first introduced in the Federal Water Act, this instrument has been increasingly strengthened through the Federal Flood Control Act 2005 and the recent reform of water legislation. Inundation zones even binding on local land-use planning and intervene in this manner into local self-government (Hartmann 2009, 2011). The recent water law does not integrate formally this well-elaborated and powerful instrument in the new plans and maps that water managers need to prepare (Rolfsen 2009: 770; Kotulla 2010: 86). The legislator has retained the traditional instruments of water management and added further instruments according to the new paradigm.

German water management agencies are now confronted with making catchment-wide comprehensive water management plans. The flood risk management plans that water management needs to prepare according to § 75 WHG are defined with respect to their aims: namely, they should reduce the negative consequences of flooding and determine appropriate measures for risk management in order to protect human beings, the environment, the cultural heritage, and the economy. The procedures and conditions for preparing such plans are not, however, determined in law. Following an analysis and anticipating the future, they have to provide a '*Vorentwurf einer normativen Ordnung*' – a preliminary sketch of a future normative order (Durner and Ludwig 2008: 458). In analysing the Water Framework Directive, Breuer identifies two possible reactions of water managers to the challenge to change their mode of governance as a result of the shift from condition-based to performance-based legislation: legal overzeal or conceptual fatalism (Breuer 2007). Legal overzeal would result in unsuitably-detailed large-scale plans; its implementation would be claimed by water managers as obligatory for the lower-level administration. Conceptual fatalism results in a situation where the administration remains unimpressed by preliminary drafts of plans as long as they are not binding. Although these are extreme positions, they demarcate the range of reaction of water managements towards flood risk management plans.

Performance-based Legislation – an Optimal Perturbation?

The European Court of Justice has decided that future water management needs to provide a coherent and comprehensive model in the form of concise and structured planning for the entire territory of the member states (Breuer 2007: 506). However, the first attempts at water management seem to move in another direction. German water management, embedded in the micro-governance of riparian systems, and based on single-case decisions (according to performance-based legislation), struggles with the development of a programmatic macro-governance of whole catchments (Breuer 2007: 508). German water management pursues a 'path of minimal institutional change' (Moss 2004: 91). Since the institutional framing does not fit in with the European water governance paradigm, German water managers have become heavily dependent on non-formalized forms of cooperation (Moss 2004: 91). So, the current relationship between water management and spatial planning seems to be losing its 'more or less stable' patterns (Wiering and Immink 2006: 424) that formed the policy arrangement between both for decades. It seems that the shift in German Water Legislation towards a performance-based regulation for flood risk managements is a breaking point in the paradigm shift towards spatial water management. A change in the policy arrangement of spatial planning and water management is pending. But the success of performance-based legislation depends 'on the ability of government agencies to specify, measure, and monitor performance'. Such reliable and appropriate information about performance maybe well-nigh impossible to obtain (Coglianese et al. 2002: 708), so new technologies are needed to cope with the requirements of performance-based water management (see Zimmermann 2010).

In reflecting on water management and spatial planning, Wiering and Immink raise the question whether the contemporary paradigm shift in these fields leads to a replacement of institutions (Wiering and Immink 2006: 423). The last reform of German water law seems to point in the direction of the coexistence of 'old, stable, and new dynamic patterns in policy concepts and institutional rules of the game' (Wiering and Immink 2006: 424). The regulations for flood protection contain not only 'old' condition-based regulations, but also 'new' performance-based regulations for property governance. The 2010 reformed Federal Water Act sticks in many respects to procedures that serve the condition-based mode of governance of water management (Lauer 2010: 637). Such changes are long-lasting social and cultural processes (Ellis and Thompson 1997: 209). Durner and Ludwig describe the struggle with performance-based water legislation as a specific German problem (Durner and Ludwig 2008: 463). A strong reliance on formal administration, procedures and the rule of law are characteristic of German water management (Moss 2004: 90). Owing to the legal traditions of performance-based and condition-based legislation, the tension between the traditional technical-oriented water management and the relatively new plan-oriented European water policy is accentuated in Germany.

That is not to say that water managers in other countries do not struggle with new paradigms in water management. Van den Brink addresses the challenges of contemporary water management in the Netherlands in her book *Rijkswaterstaat on the Horns of a Dilemma*: The dilemma is between the technical expertise and requirements of water management on the one hand, and the increasing social and political imperative on the other (van den Brink 2009). But, according to the type of legislation, the Netherlands have a better starting position for the paradigm shift. Dutch water management is a relatively autonomous layer of public administration, traditionally more used to performance-based regulations (Wiering and Immink 2006: 428). Dutch spatial planning adheres very much to formal procedures (Wiering and Immink 2006: 427). So, the legal traditions seem to be the opposite of those in Germany.

The question how different types of legislation – the condition-based and the performance-based– determine modes of governing land in different countries is crucial because of the need for international cooperation in water management. Dutch water managers, as downstream stakeholders on the River Rhine, are unlikely to understand the struggle of German water managers in spatial planning behind the dykes. Thus, it is important to understand how the struggle of German water managers and spatial planners results from performance-based and condition-based legislation.

The paradigm shift in water management and thus also in spatial planning is inevitable and irreversible. It is inevitable because contemporary (European) legislation induces the paradigm shift; in particular the European Flood Risk Management Directive pushes the shift. It is irreversible because of the technological constraints of contemporary flood protection in the face of increasing flood risk (see also Hartmann 2011). The trend towards performance-based legislation in environmental issues and the increasing environmentalization (Wegener 1999) in spatial planning demands a learning process from land and water institutions, namely that they collaborate with each other and with institutions in other countries. The German situation especially needs further research attention, because of the particularities of the types of legislation there and because of the geographical and political central role in Europe. It seems that performance-based regulations provide more flexibility, and might therefore be better suited to wicked planning problems – as described in the first chapter of this book. Condition-based regulations seem to provide more institutional stability and legal security. Further theorizing is needed about the role of performance-based and condition-based regulation for planning by legislation and property rights.

In conclusion, we have shown that when water managers are obliged to manage riverside property behind the dykes – as a consequence of spatial water management – they are planning in an unknown arena. Whenever different institutions are forced to work together in a new or uncommon arena, the question of the different modes of governance becomes crucial. This contribution explains how essential it is to recognize how the types of legislation determine the institutions' actions in managing riverside property. How institutions learn and adapt modes of governance is thus not just a theoretical planning question, but also a legal question.

Acknowledgement

Special thanks to Richard K. Norton, chair and associate professor at Michigan State University, for discussing the appropriate terms for performance-based and condition-based legislation; and to professor Willem Salet, University of Amsterdam, for his feedback.

References

Albrecht, J. and W. Wendler 2009. Koordinierte Anwendung von Wasserrahmenrichtlinie und Hochwasserrisikomanagementrichtlinie im Kontext des Planungsprozesses. *Natur und Recht*, 31(9), 608–618.

Albrecht, J. 2007. *Umweltqualitätsziele im Gewässerschutzrecht: Eine europa-, verfassungs- und verwaltungsrechtliche Untersuchung zur Umsetzung der Wasserrahmenrichtlinie am Beispiel des Freistaates Sachsen*. Berlin: Duncker and Humblot.

Ang, G., Groosman, M. and Scholten, P.M. 2005. Dutch performance-based approach to building regulations and public procurement. *Building Research and Information*, 33(2), 107–119.

Breuer, R. 2007. Praxisprobleme des deutschen Wasserrechtsnach der Umsetzung der Wasserrahmenrichtlinie. *Natur und Recht*, 29(8), 503–513.

Coglianese, C., Nash, J. and Olmstead, T. 2002. Performance-based regulation: prospects and limitations in health, safety, and environmental protection. *Administrative Law Review*, 55(4), 705–728.

Durner, W. and Ludwig, L. 2008. Paradigmenwechsel in der europäischen Umweltrechtsetzung? *Natur und Recht*, 30(7), 457–467.

Ellis, R.J. and Thompson, M. (eds) 1997. *Culture Matters: Essays in Honor of Aaron Wildavsky*. Boulder, CO: Westview Press.

Fonk, C.F. 2010. Die konditionale Rechtssetzung in der Tradition Otto Mayers: einantiquiertes Normstruktur- und Gesetzgebungsmodell? *Deutsches Verwaltungsblatt*, 10, 626–633.

Gerdes, F. 2008. *Wiederherstellung von Retentionsflächen: Fallanalyse der Retentionsflächen Köln–Langel und Köln–Worringen* (unpublished diploma thesis). Dortmund.

Glasbergen, P and Driessen, P.P. 2002. The paradigm shift in environmental politics: towards a new image of the manageable society, in P.P. Driessen and P. Glasbergen

(eds), *Greening Society: The Paradigm Shift in Dutch Environmental Politics*. Dordrecht, Boston: Kluwer Academic Publishers, 3–26.

Hartmann, T. 2009. Clumsy floodplains and the law: Towards a responsive land policy for extreme floods. *Built Environment*, 35(4), 531–544.

Hartmann, T. 2011. *Clumsy Floodplains: Responsive Land Policy for Extreme Floods*. Farnham: Ashgate.

Ison, R., Röling, N. and Watson, D. 2007. Challenges to science and society in the sustainable management and use of water: investigating the role of social learning. *Environmental Science and Policy*, 10(6), 499–511.

Knopp, G.M. 2010. *Das neue Wasserhaushaltsrecht: WHG-Novelle 2010, Gewässerbenutzung, Ausbau*. München: Beck.

Kotulla, M. 2010. Das novellierte Wasserhaushaltsgesetz. *Neue juristische Wochenzeitschrift*, 79(2), 79–86.

Krautzberger, M. (ed.) 2007. *Baugesetzbuch – Kommentar. Version 86*. München: C.H. Beck.

Lauer, K. 2010. Das WHG 2010: Weichenstellung oder Interimslösung. *Natur und Recht*, 32(9), 635–638.

LAWA 1995. *Leitlinien für einen zukunftsweisenden Hochwasserschutz*. Berlin: LAWA.

Moss, T. and Monstadt, J. 2008. *Restoring Floodplains in Europe: Policy Contexts and Project Experiences*. London: IWA Publishing.

Moss, T. (2004). The governance of land use river basins: prospects for overcoming problems of institutional interplay with the EU Water Framework Directive. *Land Use Policy*, (21), 85–94.

Moss, T. 2009. Zwischen Ökologisierung des Gewässerschutzes und Kommerzialisierung der Wasserwirtschaft: Neue Handlungsanforderungen an

Raumplanung und Regionalpolitik. *Raumforschung und Raumordnung*, 67(1), 54–68.

Mostert, E. and Junier, S.J. 2009. The European flood risk directive: challenges for research. *Hydrology and Earth System Sciences Discussions*, (6), 4961–4988.

Needham, B. 2007. *Dutch Land Use Planning: Planning and Managing Land Use in the Nederlands, the Principles and The Practice*. Den Haag: Sdu Uitgevers.

Nisipeanu, P. 2008. Tradition oder Fortentwicklung? Wasserrechtim UGB. *Natur und Recht*, 30(2), 87–97.

OVGKoblenz 2004: Hochwasserschutzauch auf landwirtschaftlich genutzten Flächen (Decision of the court), No. AZ 1 A 11787/03. [Online]. Available at: http://beck-online.beck.de [accessed 22 September 2011].

Pape, K.A. 2010. Wasserhaushaltsrecht nach der Novelle des Wasserhaushaltsgesetzes, in *Umweltrecht Kommentar. Band III*, edited by R. Landmann and G. Rohmer. München: Verlag C.H. Beck.

Paul, M. and Pfeil, J. 2006. Hochwasserschutz in der Bauleitplanung (unter besonderer Berücksichtigung des Hochwasserschutzgesetzes 2005). *Neue Zeitschrift für Verwaltungsrecht*, 25(5), 505–512.

Project F01 2006/2007 2007. *Heute schon an morgen denken: Mehr Raum für den Hochwasserschutz durch konfliktarme Flächenmobilisierung*. Final report of a student project at TU Dortmund University. Dortmund.

Project F01 2008/2009 2009. *Eine dicke Lippe riskieren: effektiver Hochwasserschutz durch innovatives Bodenmanagement*. Final report of a student project at TU Dortmund University. Dortmund.

Reinhardt, M. and Czychowski, M. 2010. *Wasserhaushaltsgesetz unter Berücksichtigung der Landeswassergesetze: Kommentar (10. edition)*. München: Beck.

Reinhardt, M. 2003. Retentionsflächen und Eigentum zur wasserrechtlichen Planfeststellung Deichverlegungen. *Zeitschrift für Wasserrecht*, 42(4), 193–212.

Reinhardt, M. 2004. Hochwasserschutz zwischen Enteigungsentschädigung und Amtshaftung. *Natur und Recht*, 26(7), 420–429.

Reinhardt, M. 2008. Der neue europäische Hochwasserschutz. *Natur und Recht*, 30(7), 468–473.

Rijksoverheid 2011. *Wet op ruimtelijke ordening*. [Online] [accessed 9 June 2011].

Rolfsen, M. 2009. Das neue Wasserhaushaltsgesetz. *Natur und Recht*, (31), 765–771.

Roth, D. and Warner, J. 2007. Flood risk uncertainty and changing river protection policy in The Netherlands: the case of 'calamity polders'. *Tijdschriftvoor Economische en Sociale Geografie*, 98(4), 519–525.

Schätzl, R. and Hoffmann, H. 2006. Agrarökonomische Untersuchungen: Teilprojekt U-7 integrated land use planning and river basin management (ILUL). [Online]. Available at: www.interreg-ilup.de [accessed 22 September 2011].

Söfker, W. 2009. *Baugesetzbuch: Textausgabe mit ausführlichem Sachverzeichnis und einer Einführung* (41st edn). München: DTV, C.H. Beck.

Spit, T. and Zoete, P. 2009. *Ruimtelijke ordening in Nederland: Een wetenschappelijke introductie in het vakgebied.* Den Haag: Sdu Uitgevers.

Steindorf, J. and Häberle, P. 2010. Gesetz zur Ordnung des Wasserhaushalts (Wasserhaushaltsgesetz – WHG): Vorbemerkungen, in G. Erbs and M. Kohlhaas (eds), *Strafrechtliche Nebengesetze* (178th edn). München: Verlag C.H. Beck.

Stüer, B 2005. *Handbuch des Bau- und Fachplanungsrechts: Planung – Genehmigung – Rechtsschutz* (3rd edn). München: Beck.

Stüer, B. 2004. Hochwasserschutz im Spannungsverhältnis zum übrigen Fachplanungsrecht, Raumordnungsrecht und zur Bauleitplanung. *Natur und Recht*, 26(7), 415–420.

Tetley, W. 1999. Mixed jurisdictions: common law vs. civil law (codified and uncodified). *Uniform Law Review*, (3), 591–618.

UBA (Umweltbundesamt) 1999. *Anforderungen des vorsorgenden Hochwasserschutzes an Raumordnung, Landes / Regionalplanung, Stadtplanung und die Umweltfachplanungen: Empfehlungen für die Weiterentwicklung* (UBA Texte No. 45/99). Berlin.

UBA (Umweltbundesamt) 2003. *Sicherung und Wiederherstellen von Hochwasserrückhalteflächen* (UBA Texte No. 34/03). Berlin.

van den Brink, M. 2009. *Rijkswaterstaat on the Horns of a Dilemma.* Delft: Eburon.

Wegener, M. 1999. Raumplanung als Systemrationalität: oder die Rettung der Raumplanung durch die Ökologie. in K.M. Schmals (ed.), *Was ist Raumplanung?* Dortmund: Dortmunder Beiträge zur Raumplanung, 89, 165–172

Wesselink, A. 2007. Flood safety in the Netherlands: The Dutch response to hurricane Katrina. *Technology in Society*, 29(2), 239–247.

Wiering, M. and Immink, I. 2006. When water management meets spatial planning: a policy-arrangement perspective. *Environment and Planning C: Government and Policy*, (26), 423–438.

Zimmermann, G. 2010. *Automatisierte Bildinterpretation: Die Zukunft der Fernerkundung in der kommunalen Planung.* Dortmund: Institut für Raumplanung.

Interviews and empirical studies were conducted between 2004 and 2008 by the author and supervised students. Interview protocols are on file with the author.

PART III
Coping with Environmental Change

Chapter 6

Improving Institutional Conditions for Adaptive Planning in the Netherlands

Joost Tennekes

From Planning for Adaptation to Adaptive Planning

Over the past years, Dutch administrative bodies for spatial planning have initiated several policy programmes that aim to make the country 'climate change proof' (National Water Plan 2009–2015, National Adaptation Strategy '*Making room for adaptation*'). Adaptation to climate change involves technical and structural measures against the threat of sea level rise, discharge waves in rivers, difficulties with urban drainage after heavy precipitation, more frequent heat waves in summer, and pressure on freshwater levels (PBL Netherlands Environmental Assessment Agency 2009, 2011; Ven et al. 2010).

To Dutch spatial planners, none of these problems are new in themselves. Spatial planning in the Rhine–Meuse–Scheldt delta has always been an exercise between accepting the natural forces that cannot be changed, and ingenious engineering to make the low-lying country safe and liveable. What is new is the determination to take an extremely long-term scope into account in their considerations. That climate change will have its effects cannot be denied, but tangible effects, let alone a breakdown of the current systems (coastal sand replenishment, dykes and freshwater supply), are still a long way off (PBL Netherlands Environmental Assessment Agency 2009; Kwadijk, Jeuken and Van Waveren 2008). Academic knowledge about the effects of climate change is being made available to planning authorities (e.g. Bessembinder 2009), but is inherently uncertain (Dessai and Van der Sluijs 2007), especially at the regional and local scales on which most interventions will have to take place. What is more, adaptation measures for the long-term future have to take into account other factors, including demographic decline and developments in employment and mobility, that are even more uncertain and at the same time will have a far greater impact on spatial structures. All in all, investments in measures to adapt to climate change run a high risk of being irrelevant to current political problems and of turning out to be disproportional.

Nevertheless, many planning decisions that are being taken now implicitly affect the possible adaptation choices for the future (Hallegatte 2009). The planning of residential areas in polders close to rivers makes it improbable that these areas could ever be designated 'inundation polders in case of a calamity'

(Werners, Warner and Roth 2010). The option of raising ground levels of residential buildings as a safeguard against rising waters exists only at the moment of construction. Planning open water or parks in residential areas is much cheaper in the initial phase than reconstructing these areas at a later stage.

So, planning decisions made in the present need to take climate change into account for the distant future. This sometimes demands new fundamental directives from planning authorities. However, research on policy implementation indicates that it is exactly the long-term character of problems of climate change that constitute the most important barrier in planning practice related to adaptation measures (Biesbroek et al. 2011)). The question examined in this contribution, therefore, is how to improve the institutional setting in such a way that more account is taken of the long-term concern for climate change in decisions by all parties involved in the planning process – whether they be planning authorities, individuals or private companies.

This question only partly concerns the availability of policy instruments that may be used for implementing adaptation measures. The existing set of policy instruments that is being applied to the more traditional policy goals in the fields of water management, spatial planning, construction and environmental protection, would be just as applicable to the objectives of adaptation (Driessen et al. 2011). Only marginal adjustments to particular instruments may be necessary, in order to increase the room for flexible and locally specific planning. Such adjustment – taking into account the time before climate change would have any effect – is not the most difficult task.

The problem is rather that adaptation measures must be taken within existing practices, considering the spatial structures and property rights already in place. The most important question, therefore, is whether the existing constellation of formal rules, institutionally tied up budgets, role attitudes and routines, and distribution of powers, are conducive to adaptation? Together, do they form the right incentive structure, so that planners would be inclined to make use of the existing policy instruments in order to achieve adaptation? This question is all the more important seeing the uncertain character of the present knowledge about local consequences of climate change. The institutional incentive structure should direct planning decisions towards adaptation, not only in the present, but should also stimulate planners in future decision situations to take into account the then available knowledge. This broadens the challenge for climate change adaptation from *planning for adaptation* that is, planning and designing of spatial structures that render the Netherlands more robust to climate change on the basis of the present knowledge, to *adaptive planning*. Adaptive planning means shaping the conditions for future planning decisions in such a way that they will take into account the best knowledge available then on changing climatic conditions.

Here, this issue is explored by addressing some important aspects for improving the institutional setting for adaptive planning. On the basis of literature research and expert interviews, suggestions for improving the institutional context – some of which originally related to other issues than adaptation – are proposed. These concern agenda setting, financing, and competencies in the public sector. In all these suggestions, the emphasis is on law as a regulative policy instrument, in addition to financial and communicative instruments.

One other important institutional aspect is the institutionally embedded distribution of risk in society. Climate change may lead to a greater responsibility for reducing risks on the part of citizens and companies than is the case in the present situation. Within the same formal setting, changes in the physical system may de facto shift responsibility to private citizens. But it is also possible that government would reassess its traditional responsibilities for weather-related risks related to climate change. Regulative instruments may be needed, not only to institutionalize this distribution of responsibilities, but also to regulate its inevitable changes in cases in which climate risk is shifted towards government or third parties.

Before analysing these aspects, it is important to have some knowledge of the actual planning issues as a consequence of climate change. The next section presents a short introduction into possible climate change effects within the Dutch urban context, and some typical technical and planning measures for counteracting them. Section 3 examines aspects of the institutional setting related to agenda setting, financing and competency conflicts. Section 4 discusses the issue of institutionally embedded risk distribution. The concluding section recapitulates the proposed institutional adjustments and their interrelatedness.

Climate Change Adaptation in the Urban Area

In general, there are three important ways in which climate change could affect Dutch cities (PBL Netherlands Environmental Assessment Agency 2009). The first is an increase in the risk of flooding to cities on the banks of major rivers or in areas that are prone to flooding by the sea. Sea level rise and greater and more fluctuating river discharges would increase the risk of a dyke breach. In the densely populated west of the Netherlands, evacuation would be difficult or near impossible if there were to be flooding by the sea, something which cannot be predicted as far in advance as can river flooding. Downstream, at locations such as Rotterdam and Dordrecht, both types of flood risks have a cumulative effect.

The second problem is that of heavy precipitation, which could become aggravated by climate change. In cities, the high building densities, surface sealing and pavements, cause a low capacity for rainwater draining into the ground, causing a large amount of run-off, which has to be handled by drainage and sewerage systems. In the nineteenth and twentieth centuries, urban expansions were built in the polders, where even under normal circumstances water has to be pumped away. More intense and fluctuating precipitation will probably lead to water nuisance such as flooded basements and streets, in turn, causing problems for traffic, city trees and parks.

Finally, more frequent and persisting heat waves could pose a health problem to the population. The well-known heat island effect means that the energy collected in the built-up city areas during the day cannot escape during the night as easily as it could in the countryside. Preventing heat collection was traditionally not as important an aspect of Dutch urban design, as it may have been in more southern regions. Increasing temperatures may lead to a situation in which the occurrence of heat waves is politically considered to be a matter of public health.

Counteracting these effects requires technical and planning measures at different scales: at the regional scale (dykes, adjusted locational choices and water buffers), at neighbourhood level, down to individual streets and buildings. For each level some examples are given below.

The first example of a measure taken at a regional scale is the 'burst-proof dyke', also called 'climate dyke'. In the low-lying Netherlands, a dyke burst could cause the flooding of huge areas of land, with water several metres deep. Different mechanisms may cause such bursts. In case of overtopping, water levels of rivers or the sea rise above the level of the dyke water runs over the top of the dyke and weakens it until it bursts (RIVM 2004. A burst-proof dyke will not burst, even when water runs over it. Because water levels will eventually decrease, only a limited amount of water spills over on to the land. The consequences in terms of expected numbers of casualties and material damage are much lower than would be the case for traditional dykes.

Burst-proof dykes may be achieved by making them very broad, up to two hundred metres. This, however, is a very costly solution, not only because of the costs involved in construction and maintenance, but also because of the value of the land, especially in areas with a high urban pressure. Combining water defence with other land functions, such as urban development on top of this broad dyke, would recover those costs and enrich the urban environment with a new milieu: apartments with a view of the water instead of the dyke (Hartog et al. 2009).

An example of a measure of a less technical nature, at a regional level, would be to adjust the locational choice for urban expansion. In the Netherlands, there has been much discussion about the suitability of deep polders for urban expansion.

When there are no alternative locations, or in the case of restructuring of old parts of a city, adaptation measures at the neighbourhood level would entail a different design at the construction site itself (Pieterse et al. 2010), for example, by raising the ground level of buildings in order to prevent water nuisance and to lessen the effects of a flood. A bigger part of these locations should be reserved for open water and greenery, such as ponds and parks, in order to reduce water nuisance and heat stress.

At smaller scales, adaptation could entail designing different street profiles and sewerage systems that could cope with more rainwater. A new solution within the Dutch context is that of 'water squares', squares that would serve as storage areas for water, but only in times of heavy rainfall (Greef 2005, Boer 2010). At the scale of individual buildings, adaptation could mean the introduction of 'green roofs';

buildings covered with a layer of earth that would not only absorb rainwater and delay run-off, but in summer would also cool the apartments beneath.

Crucial from a perspective of long-term planning are the differences in flexibility between these measures (Hallegatte 2009; Meuleman and In't Veld 2009). Street pavements are being renewed at maintenance sequences of 10 or 20 years. Therefore, city authorities may react to changes in weather patterns relatively quickly, for example, by raising footpaths. In contrast, replacing sewerage systems is done once every 50 years. Upgrading a normal dyke to become a broad, burst-proof dyke entails a complex and costly restructuring of parts of the city, and will occur only sporadically. And, finally, it is very improbable, historically speaking, that neighbourhoods will be 'relocated', in the next centuries.

The flexibility of a measure is not only determined by sequences of maintenance and replacement, but also by the public and private nature of stakeholders. Measures which exclusively affect the public domain, such as adapting the level of road surfaces, are much more easily realized by public initiatives than measures that require citizen involvement. Green roofs for all buildings cannot be decreed without infringing on the principles of good governance. Moreover, not all private owners share the same considerations when it comes to investing in their real estate. Private companies, thinking in financial terms, are likely to take measures if they would be profitable. Private owners usually are more emotionally attached to their property, and may not be inclined to take measures that would physically alter the building.

In short, in the context of existing urban structures, the implementation of some adaptation measures is more inflexible than others. Such measures have to be taken at opportune moments – mostly for other reasons than climate adaptation. An example is the extensive restructuring of high-rise neighbourhoods, built in the Netherlands in the 1960s and 1970s, which are to be restructured over the next twenty years. Although the effects of climate change may not reveal themselves within that period, adaptive planning requires that authorities take this into account now (PBL 2011).

Institutional Settings: Political Agendas, Financing, Competencies, Risk Distribution

So, technical measures against adverse effects of climate change are available, and the existing repertoire of policy instruments is sufficiently appropriate for implementing these measures. This leaves the question of whether the new issue of climate change 'fits' within current practices, in the sense that actors involved in the planning process will actually make use of the knowledge of innovative adaptation measures and existing instruments. This section addresses three ways in which existing practice could impede effective investments in adaptation: political agenda-setting processes, the practice of cost recovery in real estate development, and the conflicts over competencies of different government agencies.

Agenda Setting

The consequences of climate change will reveal themselves in the distant future, and are relevant on a time scale that is far beyond the usual political four-year horizon, but also beyond those of planning authorities, which produce plans on a time scale of ten to twenty years. Climate change may be a theme now in some municipalities, but it is not at all certain that this will stay on the agenda if the problematical effects do not show up for a long time, and other urgent and important issues push it aside (Vromraad 2007). This is not a problem when the necessary measures are of the more flexible type that may still be implemented at the moment that effects of climate change present themselves to citizens and voters. However, most of these measures involve inflexible, long-cycle decisions, which may have to be taken at a time when climate change may not have political attention. Without political attention, adaptation measures may not be included in urban development plans, or they may constitute such a 'soft' municipal interest that they will be dropped when they pose an obstacle to the realization of such plans. For the issue of adaptation not only has to compete with other public interest issues on the political agenda, such as municipal schools and swimming pools, but above all with the considerable political interest which municipalities usually have in the realization of urban development projects. Apart from the financial interests of many Dutch municipalities in urban development projects (Segeren 2007), political prestige is always at stake. Failure of the project would be very costly, both financially and politically.

New policy instruments could improve the conditions for political and administrative awareness of climate change in decision making. A first step may include the obligation of municipalities to regularly survey the local climate change effects that are expected for the future. Such an obligation already exists, for example, in the United Kingdom, under the Climate Change Bill (Swart 2009). A *climate change assessment* would take this one step further, and would entail the obligation to incorporate in the spatial planning process an assessment of the robustness of urban development plans, in the light of future climate changes. Rather than creating a new assessing instrument for this, it would be better to incorporate an assessment in existing procedures. In the Dutch context, the Environmental Impact Assessment, and the obligatory Water Assessment would qualify for this (Leusink and Zanting 2008).

There is a risk that such new assessments become just an administrative hurdle, a step that can be taken by ticking some boxes. It is important, therefore, that political significance is attached to the assessments. This could be a reason to prefer integrating long-term climate concerns into the Environmental Impact Assessment (EIA) rather than other alternatives, such as into the Water Assessment. Legal possibilities of raising objections within the EIA procedure are used by citizens or NGOs – some would say opportunistically – to oppose building projects, even when the actual reason for their opposition is not environment-related. If criteria related to climate change were to be part of this kind of political reasoning, adaptation concerns may gain more political attention than if these criteria were part of the Water Assessment, which, as we will see, is a procedure for coordination between governmental bodies only.

Financing Adaptation

Adaptation measures will entail additional costs, although they can be reduced if taken into account in spatial planning and design from the start (Pijnappels and Sedee 2010). Even small additional costs may be a reason for not implementing them. Government subsidy schemes are a suboptimal solution here. Investors could be frightened off by the possibility of future changes to subsidy regulations. It would be crucial to improve the structural possibilities for investors to recover or earn back their investment in adaptation measures.

Take for example the situation for real estate development in residential areas, where private investors and municipalities cooperate and negotiate. Under the Dutch cost-recovery act (*Grondexploitatiewet*) of 2008, participating investors, both private companies and participating municipalities, may enter into a contract ('*exploitatieovereenkomst*'; a contract between all investors under private law) about cost settlement. When agreement under private law cannot be reached, the municipality may establish a cost settlement plan under public law. In most cases, development takes place under a private law contract. However, the possibility of municipal interference in the form of binding cost settlements under public law casts its shadow over negotiations about what should be included in the private contract. So public law is important. Under current public law, municipalities could impose a mandatory inclusion of adaptation measures in the public development plan (*exploitatieplan*).

Having this legal option, however, does not automatically mean that municipalities will exercise this, in actual practice (Dinteren and Van der Krabben 2010). As discussed in the previous section, they have been found to make use of this option only when it is important to them politically. Nevertheless, in cases when municipalities *are* determined to integrate adaptation measures into their new city expansion or restructuring plans, they still have to deal with project developers who do not necessarily share their objectives. In the day-to-day practice of the Dutch land and residential real estate market, there is the problem of the 'split incentive'; investing parties are not those that benefit from investments. When project developers have shared in the costs of adaptation measures, there is no guarantee that their investments will be reflected in a higher market price. Better adaptive qualities may not be reflected in consumers being willing to pay higher prices. Even when adaptation measures would be proven money savers (as is the case with heat insulation measures), in the Dutch market new measures will not automatically lead to better prices (Hoppe and Faber 2011). Rather than prices being determined by consumer preferences about the physical characteristics of the house, they are determined to a greater extent by location and buyers' credit possibilities (Besseling 2008), or (in the regulated rental sector) by the administratively reality of given 'quality indicators'.[1]

Therefore, investors run a high risk of not being able to recover their investment in climate adaptation measures. In the Dutch situation, project developers, compared to municipalities, have gained much bargaining power over the last twenty years (Segeren 2007), and now the post-credit-crisis situation may further advance their arguments against such adaptation measures, because developers could claim that extra costs would ruin the entire development project. Thus, in the interest of adaptation, it is important to create policy instruments that support municipalities in these negotiations.

Two types of policy instruments could be relevant here. The first is the introduction of adaptation standards in national building regulations (*bouwbesluit*). As long as municipal demands with regard to adaptation are *supplementary* to existing national stipulations (*bovenwettelijk*), they are bound to be the first to disappear in the negotiations with project developers. Including such demands in the national regulations prevents this, and also may create a level playing field in the competition between project developers. A second, more geographically specific way to strengthen the position of municipalities in the bargaining process is to stimulate municipalities to formulate a generic cost recovery policy with regard to adaptation measures in their territory. As Munoz Gielen (2010) proposes, a generic recovery policy may provide more certainty to project developers with regard to their profit calculation – and following Ricardian land value theory, they would presumably pay lower prices on the land market. Not only would a generic recovery policy mean that, in effect, land owners pay for the adaptation measures, an officially documented and politically sanctioned commitment to certain adaptation demands would also help to prevent municipalities from abandoning their adaptation demands in the negotiations with project developers.

Competences

As we have seen above, the challenge to keep climate adaptation on the agenda in the process of spatial planning and design depends on the positions of power between the parties involved. This applies to more than the relationship between government bodies and private companies. The struggle over competences between government bodies is equally important. The division of responsibilities between municipalities and water boards, and their respective role attitudes, are of great importance in planning for adaptation (Neuvel 2009; Buuren and Eshuis 2010).

Dutch water boards are democratic bodies, with their own elections and powers, but unlike the municipalities, which have very wide responsibilities,

---

1 The split incentive and the resulting uncertainty about the investment returns do not equally apply to all real estate: 'green' offices have been reported to realize higher prices (Brounen and Kok, 2009).

they are functional government bodies, with a specific responsibility for water management. This may appear to be only a functional division of responsibilities, but in actual practice it is a territorial one, as well; water boards are expected to take the lead in policy making for rural areas and with regard to dykes (which are kept free from building development). Their interference in the development of built-up areas is much smaller, but they do facilitate it, by seeing to it that these areas stay dry.

This division of responsibilities causes problems in cases of city expansions at locations that can be expected to be confronted with problems of drainage, as would be the case in some deep polders. Given climate change, these locations are considered suboptimal. However, Dutch municipalities tend to expect the water boards to solve these problems by taking technical measures inside as well as outside the borders of such development sites. Although this is technically possible, in this way adaptation costs are shifted onto the water boards. A couple of years ago, Water Assessments were introduced to solve these problems. Municipalities have to seek the advice from water boards on all development plans. In theory, water boards may advise against a proposed location, and have some legal means to enforce their advice. In actual practice, however, water boards do not use these means to the full (Dutch Ministry of Transport, Public Works and Water Management 2006. Water boards are reluctant to enforce negative advice on locational choice, for fear of harming the relationship with the municipalities. One of the reasons for this situation is that they are dependent on municipal cooperation for their own projects (for example for rural projects to broaden drainage canals).

Therefore, water nuisance risks are addressed not by changing the location but by adapting the design, for example, the inclusion of more open water, such as ponds and canals, into development plans. But then, too, differences in roles and attitudes are obstacles to the cooperation between municipalities and water boards. Municipalities complain that water board employees do not treat water as an integral part of the design of the plan, and that they approach the problem 'technically', only creating 'engineer's water', instead of 'water as an amenity', such as in the form of water gardens. In contrast, municipalities see themselves in the role of 'urban developers', creating residential areas in which water adds value to the quality of life. According to them, water board employees merely 'inspect', and are only interested in controlling their standards with regard to water, whereas municipality employees are interested in the resulting, overall quality of the urban design.

To counteract this sectoral divide, there are 'soft' policy instruments, such as the '*functiefaciliteringskaart*', a map mentioned in the *National Water Plan*, with information on the spatial functions that water could perform, but it is questionable whether these instruments are sufficient or not. Maybe water boards could be made structurally more independent of municipalities by granting them more autonomy in implementing their own projects (Franssen and Groeneveld 2009).

The existing distribution of powers may stand in the way of innovative design, where government parties are reluctant to abandon their routines and role perceptions. The case of the climate dyke (burst-proof dyke) illustrates this point. In the Netherlands, policy regarding the most important defences against flooding (what, in Dutch, is called the *primaire keringen*) is in the hands of a specialized service of the Dutch Ministry, the Directorate-General for Public Works and Water Management (Rijkswaterstaat), has the task of designing the policy for all the important dykes, dams and dunes, whereas the water boards are involved in the construction, monitoring and maintenance of these water barriers. Rijkswaterstaat has a long history in Dutch public administration, and a huge status in terms of having the technical competency to protect the Dutch population against flooding. Institutionally, the system is set up to ensure that Rijkswaterstaat may execute its tasks without interference. The budget is set nationally, and recent policy reforms aim to create long-term funding for the nation's water defences, so that it will no longer be subject to the political budgeting of future government coalitions. The construction of water defences is also institutionally separated from the spatial planning system. In the spatial planning system, the decision-making process is mainly concentrated on the local municipal level. However, the water defence works themselves, and a zone around them, generally have to be kept free from urban development, so that they may be inspected, maintained, and strengthened when necessary. With the aid of project plans, Rijkswaterstaat is able to incorporate the construction of their water defence systems into local spatial planning.

In this institutional context, climate (burst-proof) dykes, as described in Section 2, are a problematic issue. Such dykes provide a greater security, but their charm (and the ability to recoup some of the huge extra investment costs) is to be found in the fact that they could be used for more purposes than water defence only (Hartog et al. 2009). These dykes would be built on, and maybe there should also be constructions within the body of the dyke, such as car parks or transport systems. This, in turn, would imply that Rijkswaterstaat and the water boards would lose their protection against interference, in various ways. The first interference would take place at the moment of construction of the dyke. The budget for construction would become subject to local decision making and local dynamics on the real estate market. Private parties, such as developers, would have much influence on the project design. Rijkswaterstaat and the water boards would cease to be able to decide, by themselves, when and how these dykes would be build. They would – far more than is the case today – have to take local developments into account. In second place, inspection of these dykes would become more difficult, as they would be covered by buildings and pavements and such. Moreover, the real estate investors involved would need some assurance that their property would not have to be torn down for new construction work on the dyke, for a reasonably long period of time. This would conflict with the five-year cycle of Rijkswaterstaat to reassess the strength of dykes.

In the Dutch city of Tiel, next to the river Waal, an answer to these problems is found in the design of the dyke. Rijkswaterstaat has calculated what the 'theoretical' profile of the dyke should be, in order for it to meet the standards. In the total body of the dyke, this profile should not be affected by any building construction. The rest of the dyke can be developed according to the wishes of other actors.

The design of the climate dyke at the Dutch town of Tiel may be interpreted as an elegant solution to (a part of) the institutional problem. At the same time, it is an illustration of the reluctance of the parties involved to break out of the existing division of competences, and to find new collaborative arrangements for designing innovative dykes, where building houses at the waterfront does not compromise water safety procedures.

Distribution of Climate Change Risk as an Institution

Adaptive planning takes place in interaction between government bodies, between government and citizens, and between government and private sector companies. This interaction is more likely to result in commitment and investment in adaptation measures, if parties involved have a clear and reliable idea of the political choices with regard to the distribution of risks, costs, benefits, and responsibilities. To private sector actors and citizens, the limits of government action for adaptation should be crystal clear, in two different respects First, government should communicate about when it would not be responsible for taking measures to prevent negative effects of climate change. Second, government should specify to what extent it is not liable when adaptation measures are unable to prevent any damage, in spite of government action. The first is a matter of taking responsibility politically, the second of the distribution of risk and liability, and thus of taking legal responsibility.[2] Both aspects of limited government responsibility leave

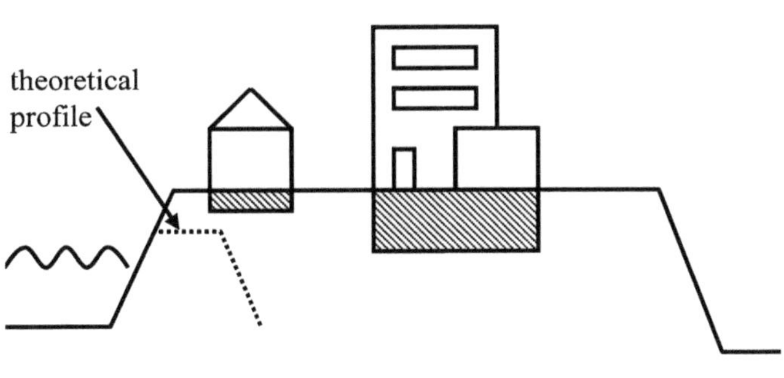

Figure 6.1 Indicative sketch of the design of the climate dyke at the Dutch town of Tiel

2 Of course both aspects are interrelated: in cases where government does not act according to its own standards to prevent damage, it may be liable.

adaptation to private stakeholders, including the decisions between taking damage preventing measures and accepting the risk.

The traditional distribution of weather-related risk is institutionally embedded in jurisprudence and taken-for-granted expectations towards government, rendered permanent in organizations and procedures. Fundamental changes in climate require that this distribution be reviewed politically – a process which may or may not have the same outcome as the traditional distribution (Driessen and Van Rijswick forthcoming). Rather than giving definitive answers to this normative question – which is the subject of the discourse in political philosophy, economic welfare theory, or constitutional law – what follows in this section illustrates the fact that adaptive planning calls for institutionalization of these normative choices for certain policy instruments. New policy instruments may be necessary, both in order to institutionalize a new distribution of risk, and to regulate the cases where this risk is shifted onto other stakeholders.

The legal basis for state intervention related to adaptation may be found in European and Dutch constitutional law (Driessen et al. 2011). The latter gives the government a responsibility to ensure the habitability of its territory and to protect the local environment. This means that the government is legally obliged to protect the interests of citizens, such as in economic development, safety against flooding, or a healthy environment. However, it does not mean that citizens can demand these things from their government as a right. But as these government responsibilities are stated in a rather abstract way in the constitution, it is not clear how far they extend. Which level of risk to citizens could be considered 'normal'? The fact that government considers certain kinds of risks as its responsibility does not mean that it shields against all damages, and under all circumstances. Government supplies a minimum level of protection against risk, varying according to the issue. For protection against life-threatening dyke breaches, the minimum level of protection is probably higher than for water 'nuisance' or water shortage, both non-life-threatening events. Some events remain just 'bad luck' or a business risk. Communicating this risk distribution contributes to realism about what may be expected from government. It maintains the self-regulating capacity of society (Van Rijswick and Salet 2010) and the citizens' capacity to improvise (Gupta et al. 2010). Government solutions for climate change problems should not stifle its citizens' awareness of problems. People have to be made aware of their own contribution to solving the problem, as well as of the fact that they may have to deal with difficult situations if government is not able to prevent this.

In the urban context, a variety of examples may be found. One example is the discussion about government responsibility for city expansions not protected behind dykes, for example, old harbour complexes in cities such as Rotterdam and Dordrecht. Is government obliged to implement measures to guarantee safety against flooding, or can it let citizens live there at their own risk? To what extent should the government take measures not only to save lives, but also to prevent material damage, given the fact that, in the Netherlands, it is not possible to be privately insured against flood damage (Rosenthal and 't Hart 1998)? In the current situation, the government is hesitant to give permission for activities outside the dyke, exactly because it feels responsible, and therefore blocks initiatives in these areas. Pilot projects with these legal issues in the context of Heijplaat (Rotterdam) form a positive exception, which could provide more clarity in the future.

Another example is the concept of the duty of the municipal government to manage the disposal of rainwater. Municipalities have the obligation to carefully design sewerage system capacity, in order to handle heavy rainfall. However, for the sake of efficiency, it is not obliged to design a system that could handle the most heavy rain storms possible. Provided that sewerage systems for rainwater and waste water are separated, municipalities can allow additional drainage of rainwater onto public streets. As long as street surfaces have been designed according to the required standards, government is not liable for damage to private households caused by incoming water from the street. House owners are themselves responsible for knowing that their house is located at a low point in the street (which, in the Dutch context of a flat country, is not evident at first sight), as well as for taking the necessary measures. With uncertain knowledge about future precipitation levels, the question is when government could be considered to have taken future changes sufficiently into account with respect to its 'duty to manage the disposal of rainwater'.

Even when no formal changes are made in the distribution of risk between government and society, climate change may shift responsibility onto citizens. Take for example the right to drain rainwater onto the public streets (Bergsma et al. 2009). Under the present legislation, the rights and responsibilities of citizens are reasonably clear: in principle, property owners are obliged to dispose of rainwater themselves, not draining it onto the public domain – as far as this is reasonable (this last clause means that owners of an apartment cannot be obliged, because they do not have any land at their disposal to discard water on to). Only under circumstances of *exceptionally* heavy rainfall may property owners drain the water onto the street. Since climate change is likely to entail more heavy rainfall, such situations cease to be exceptional, and in this way there will be a shift towards more formal responsibility for private owners, who will have to discharge more water than before. In actual practice, however, in cases of water nuisance, municipalities step in, giving 'the residents the impression that it is the municipality's responsibility and they are covering up for past lapses. Furthermore, it reinforces their belief that the state should be held responsible for addressing these issues, especially as payments in taxes are made to the state for water. This contradicts the recent trend set by the Dutch government to emphasize individual responsibilities in local water management.' (Bergsma et al. 2009 45–46).

Marking out the responsibilities between the different actors in society also defines the cases where risks are unjustly passed on to others or on to government. Delimiting the responsibilities fixes a kind of 'starting point': which risk 'originally' belongs to which actor? Of course, this distributive decision is a highly political one. When a river is slowed down by 'renaturalizing' the bed upstream in order to prevent future water nuisance in the city downstream, and that causes damage to the bordering farmland, is the city passing on its problems to the farmers upstream? Or is this just a correction of the historical mistake of straightening riverbeds in order to gain farmland, something which caused the problems of cities downstream in the first place? When a property developer in cooperation with the municipality creates a new residential area and occupies buffer capacity that could have been used by the water board to store water in situations of water nuisance, is this 'normal procedure' or is it a case of 'passing the buck'?

In Dutch water policy, the fact that it is prohibited to pass on the risk, constitutes an important principle. But in actual practice, these situations cannot be prevented. Improving institutional conditions for adaptive planning includes creating instruments through which the 'the buck' may be passed in a *regulated* way, for example, by compensatory systems. Some propose private contracts between municipality and water boards, according to which developers would pay the water board a set amount per square metre of developed land. Alternatively, with a 'banking system for water storage capacity', municipalities can compensate loss of water storage capacity in one part of the city with capacity surplus in the other (Novioconsult 2010).

Climate change requires a fundamental and political discussion about the distribution of risk within society. When the government leaves no doubt about the limits of its adaptation efforts and about the rules that apply to situations where risk is passed on, private actors can plan their investments more efficiently.

Conclusion

Throughout the centuries, increasingly efficient technical and institutional systems have made it possible for people to live safely within the Dutch delta. They were so successful, that planning and water safety have been more or less separated as two distinct issues. The prospect of climate change requires that these issues become integrated once again. Adaptation to climate change involves not only technical measures, such as higher dykes and large-capacity sewerage systems, but it also requires rethinking the relationship between natural risks and planning. Although the political tide in the Netherlands is not always favourable for raising the issue of climate change, within the community of planners – at ministries, municipalities and planning engineering consulting firms – adaptation to long-term climate change has become established as an important public interest in planning and design, at all spatial scales. Adaptation is even welcomed by the planning community, as it gives a new impulse to innovative thinking, exploring new ways for 'living with water'.

However, because of the uncertainties attached to forecasting developments on the time scale of fifty to hundred years, the challenge lies not in the *planning for adaptation*, but in *adaptive planning*. The possibility of knowing exactly what effects climate change will have in a far-away future has its limits – especially at the scale on which many adaptation measures must be taken: the local scale.

Although for certain changes the direction is clear (more and more fluctuating river discharges, more fluctuating precipitation, and higher sea levels) massive investments now in adaptation measures carry a risk of overinvestment. Adaptive planning requires a planning system that takes into account the newest insights into possible climate change at the very moment a planning decision must be taken, and not a once-and-for-all decision taken a long time in advance. The decentralized planning system of the Netherlands requires thatlocal decision makers have the incentives to take into account the robustness of their plans in the light of climate change. This broadens the adaptation agenda for the Dutch Central Government. Apart from some central decisions about the physical water system at a national scale (for example about the discharge capacity of rivers), the challenge is to adjust the planning system, so that local planners do not shift possible climate change problems on to future generations. Adaptive planning moves climate issues into planning decisions by creating the institutional structure that stimulates stakeholders to take into account available knowledge about climate change in every important planning decision.

This makes the adaptation agenda one of institutional design. In the preceding sections, this agenda has been explored. Four objectives have been identified, by analysing the weaknesses of the present institutional setting from the perspective of long-term adaptive planning, and by giving some examples of concrete solutions that are being proposed. *Agenda setting* for a long-term interest in a fluid and short-sighted political environment is more feasible when there are procedures for evaluating the robustness of plans against possible climate change, and when political antagonists may point to weaknesses in this evaluation. *Finding investors* for adaptation measures in a market environment where customers may not fully pay for the increased safety against negative climate change effects would be less difficult if these investors were confronted with a stable set of (municipal) adaptation requirements. This would enable them to take any costs into account from the start. Investors may be offered possibilities to earn back their investments in other ways, for example, by getting paid for 'water services' that are now provided by water boards. Existing *struggles about competences* between public sector organizations may impede innovative thinking by separating territories, where the best solutions may be found by overlapping rationalities of planning and technical measures. Fourth and finally, the *distribution of responsibilities* needs to be addressed. Part of the right institutional setting is a clear communication from government about the limit to its responsibility for the risk, and about when societal actors carry responsibility themselves. Climate change may present an opportunity to change these limits. Where the risk is passed on to the government or to others, compensation mechanisms may be created by government. A clear and stable distribution of risks and responsibilities would support private adaptation initiatives.

In order to improve these four aspects of the institutional setting, attention is needed for adjusting existing policy instruments (for example the environmental impact assessment to incorporate a climate assessment), inventing new ones (for example the municipal generic cost recovery policy, or the 'banking system for water buffering capacity'), or altering existing power relations between organizations in the field (water boards vis-à-vis municipalities).

Of course these four policy aspects are not mutually exclusive, and they have to be implemented so as to reinforce each other. For example, producing a clear analysis of all the cases of violation of the principle that risk – and the costs of preventing that risk – should not be passed on to the other actors may produce a political issue, creating an opportunity for bringing the issue of climate change on to the agenda. Systems of compensation, in cases where risks are passed on, help to find funding when market forces will not pay without the obligation to compensate. And where municipal government draws up a generic policy for necessary adaptation measures in real estate development, this would give a clear signal about what the government regards as its responsibility according to minimum standards and about what would be left to individual responsibility.

Almost none of the issues are totally new. They have been raised before in the context of discussions on water safety, or on the quality of public space in new urban development areas. The new theme of adaptation to climate change only gives a window of opportunity to change existing constellations for the better.

Moreover, climate change is not unique in its need for an adaptive approach in planning. In this contribution, relatively specific suggestions have been made so as to improve institutional conditions. These suggestions are 'partisan' in that they are explicitly directed at changing the institutional conditions in the interest of some, mostly water-related problems. In that sense, the concept of adaptive planning as elaborated here is only applicable to climate change.

At the same time, the issue of planning for an uncertain future is a more general planning dilemma. Other major developments in the distant future are at least as important for present planning decisions as climate change. The consequences of demographic decline for spatial planning, for example, are not easily predictable given the many uncertainties about future immigration, new techniques of communication and mobility, and other societal changes. The concept of adaptive planning – using institutional design in order to direct planning decisions to taking into account of available knowledge – may be applicable to these issues, too. The four different aspects of agenda setting, investment conditions, competences, and a reassessment of the distribution of responsibilities between state and society give a broad outline of the kind of issues that have to be addressed.

References

Bergsma, E., Gupta, J. and Jong, P. 2009. *Institutions for Climate Change, Case Study on Individual Responsibility in Adaptive Capacity*, Working document 7. Amsterdam: IVM Institute for Environmental Studies.

Besseling, P, Bovenberg, L., Romijn, G. and Vermeulen, W. 2008. De Nederlandse woningmarkt en overheidsbeleid: over aanbodrestricties en vraagsubsidies,

in F.J.H. Don (ed.), *Agenda voor de woningmarkt*. Amsterdam, Koninklijke Vereniging voor de Staatshuishoudkunde.
Bessembinder 2009. *Klimaatschetsboek Nederland: het huidige en toekomstige klimaat, report no. 223*. Bilthoven: KNMI.
Biesbroek, G.R., Klostermann, J.E.M, Termeer, C.J.A.M., and Kabat, P. 2011. Barriers to Climate Adaptation in the Netherlands. *Climate Law* (2), 181–199.
Boer, F. 2010. Watersquares, the elegant way of buffering rainwater in cities. *Topos* (70), 42–47.
Brounen, D. and Kok, N. 2009. *On The Economics of Energy Labels in the Housing Market, Working Paper*. Maastricht/Rotterdam: Erasmus University Rotterdam/Maastricht University.
Buuren, A. van, and Eshuis, J. 2010. Knowledge governance: complementing hierarchies, networks and markets?, in R. In't Veld (ed.), *Towards Knowledge Democracy*. Heidelberg: Springer.
Dinteren, J. van and der Krabben, E. van 2010. Effecten Grondexploitatiewet. *Openbaar bestuur*, 1, 10–14.
Driessen, P.P.J., de Gier, A.A.J., Meijerink, S.V., Pot, W., Reudink, M.A., van Rijswick, H.F.M.W., Schueler, B.J., Tennekes, J. and Termeer, C.J.A.M. 2011. *Beleids- en rechtswetenschappelijke aspecten van klimaatadaptatie*. Wageningen: Kennis voor klimaat/Klimaat voor ruimte.
Driessen, P.P.J. and Van Rijswick, H.F.M.W. [forthcoming]. Normative Aspects of Climate-Adaptation Policies. *Climate Law*.
Franssen, R. and Roeleveld, G. 2009. *Ruimtelijke strategie, uitwerking Nationaal Waterplan, deelproject instrumenten*, rapportnummer 1201581-002. Utrecht: Deltares.
Greef, P. de et al. 2005. *Rotterdam Waterstad 2035*. Rotterdam: Episode Publishers.
Gupta, J., Termeer, C., Klostermann, J., Meijerink, S., van den Brink, M., de Jong, P., Nooteboom, S. and Bergsma, E. 2010. The adaptive capacity wheel: A method to assess the inherent characteristics of institutions to enable the adaptive capacity of society. *Environmental Science and Policy*, (13), 459–471.
Hallegatte, S. 2009. Strategies to adapt to an uncertain climate change. *Global Environmental Change*, (19), 240–247.
Hartog, M. et al. 2009. *Klimaatdijk, een verkenning*. Wageningen: Kennis voor Klimaat.
Hoppe, T. and Faber, A. 2011. Waarom de transitie van de woningsector naar een duurzame energiehuishouding niet opschiet, een expertonderzoek naar barrière. *Milieu Dossie*r (17) 4, 21–25.
Kwadijk, J., Jeuken, A., and van Waveren, H. 2008. *De klimaatbestendigheid van Nederland Waterland, verkenning van knikpunten in beheer en beleid voor het hoofdwatersysteem*. Delft: Deltares.
Leusink A. and Zanting, H.A. 2008. *Naar een klimaatbestendig Nederland, kaders voor afweging*,definitiestudie fase 1. Wageningen: Klimaat voor Ruimte.
Meuleman, L. and In't Veld, R. 2009. *Sustainable Development and the Governance of Long-term Decisions*, RMNO-series Preliminary studeis nr.

V.17. The Hague, Advisory council for research on spatial planning, nature and the environment.

Ministerie van Verkeer en Waterstaat. 2006. *Watertoetsproces op weg naar bestemming. Landelijke evaluatie watertoets*. Den Haag: Ministeren van VenW.

Munoz Gielen, D. 2010. *Capturing Value Increase in Urban Development*. Leiden: Sidestone Press.

Neuvel, J. 2009. *Geographical Dimensions of Risk Management, the Contribution of Spatial Planning and Geo-Ict to Risk Reduction*. Wageningen: Wageningen University.

Novioconsult 2010. *Betalen voor klimaatmaatregelen, essay over financieel instrumentarium Klimaatadaptatie*. Nijmegen: Novioconsult.

PBL Netherlands Environmental Assessment Agency. 2009. *Wegen naar klimaatbestendig Nederland*. The Hague/Bilthoven: Netherlands Environmental Assessment Agency.

PBL Netherlands Environmental Assessment Agency. 2011. *Een delta in beweging, bouwstenen voor een klimaatbestendige ontwikkeling van Nederland*. The Hague/Bilthoven: Netherlands Environmental Assessment Agency.

Pieterse, N., Knoop, J., Nabieliek, K., Pols, L. and Tennekes, J. 2010. *Overstromingsrisicozonering in Nederland*. The Hague: PBL Netherlands Environmental Assessment Agency.

Pijnappels, M.H.J. and Sedee, A.G.J. 2010. *Klimaat als kans*. Wageningen: Programmabureau Kennis voor Klimaat.

Rijswick, M. van and Salet, W. 2010. Een strategish kompas voor duurzame beheersing van klimaatvraagstukken. *Beleid en Maatschappij*, (37)1, 43–58.

RIVM 2004. *Risico's in bedijkte termen, een evaluatie van het beleid inzake de veiligheid tegen overstromignen*, report 500799002. Bilthoven: RIVM.

Rosenthal, U. and 't Hart, P. 1998. *Flood Response and Crisis Management in Western-Europe*. Berlin: Springer Verlag.

Segeren, A. 2007. *De grondmarkt voor woningbouwlocaties, belangen en strategieen van grondeigenaren*. The Hague: Netherlands Institute of Spatial Research.

Swart, R., Biesbroek, R., Binnerup, S., Carter, T.R., Cowan, C., Henrichs, T., Loquen, S., Mela, H., Morecroft, M., Reese, M. and Rey, D. 2009. *Europe Adapts to Climate Change, Comparing National Adaptation Strategies*. Helsinki: PEER.

Ven, F van de, van Nieuwkerk, E., Stone, K., Zevenbergen, C., Veerbeek, W., Rijke, J. and van Herk, S. 2010. *Building the Netherlands Climate Proof: Urban Areas*. 1201082-000-VEB-0003, Delft/Utrecht: Unesco-IHE/ Deltares.

Vromraad 2007. *De hype voorbij: klimaatverandering als structureel ruimtelijk vraagstuk*. The Hague: Vromraad.

Werners, S., Warner, J. and Roth, D. 2010. Opponents and supporters of water policy change in the Netherlands and Hungary. *Water Alternatives*, 3(1), 26–47.

Chapter 7
Global Climate Change and the Stability of Property Rights

A. Dan Tarlock

Global climate change will impact the use and enjoyment of land. There has been considerable research on its likely physical and social impacts, but there has been little effort to examine the impact that these changes will have on the law of property entitlements. Historically, property law – in civil, common, and mixed systems – consists of a set of abstract, universal principles that can be adapted to all changes in the ways in which societies use land. The primary functions of the law have been to promote title security and alienability subject to minimal constraints. Both functions are secured by laws that recognize exclusive entitlements designed to decrease the risk that other private parties or the state can disturb an owner's title. In contrast, global climate changes introduces new risks to property holders such as boundary shifts between private and public ownership, the steep loss of value from 'Acts of God' rather than state regulation, and new pressures for state regulation to adapt to a changing climate. Climate change will also benefit many land owners. For example, land may be more desirable for human settlement and crop production, but the consequences of this for property law and regulation are beyond the scope of this article.

This chapter will examine two issues. First, it will discuss existing examples of property rights which have a built-in expectation of change such as water boundaries, in order to speculate about the lessons they teach for incorporating the risks of climate change into the law of property. Second, it will explore how global climate change may introduce the concept of moral hazard into the duty of the state to compensate victims of excessive regulation. The law often encourages property owners to develop in the face of known natural hazards because property owners know that either the state cannot force them to avoid the hazard or that they will be compensated if the hazard occurs and destroys their property. Global climate change may provide a new and more legitimate and powerful justification for regulations which compel landowners to avoid activities that put their property at risk.

Global Climate Change, Terrestrial Impacts and Property Rights

The potential impacts of global climate change on the use and enjoyment of land are becoming increasingly better understood. However, there has been little effort to examine the consequences of these potential impacts on the private law of property entitlements. Historically, property law in civil, common and mixed systems consists of a set of abstract, universal principles that can be adapted to any changes in the ways in which societies enjoy and transfer interests in land. The major function of the law has been to promote title security by decreasing the risk that other private parties or the state can disturb an owner's title. In contrast to the status quo maintenance bias in property, global climate change introduces new risks to property holders such as boundary shifts between private and public ownership, the steep loss of value from 'Acts of God' rather than state regulation, and new pressures for state regulation of land to increase the moral hazards of ownership. The chapter first examines selected existing examples of property rights which have a built-in expectation of change such as water boundaries, to provide some models for global climate change-influenced rules. It next examines the incorporation of the moral hazard concept into property protection law to speculate how the risks of climate change can be incorporated into the law of property in order to provide a rationale for increased state regulatory power to deal with global climate change.

There are two strategies to address the projected adverse impacts of global climate change: mitigation and adaptation. Most domestic and international efforts have been directed toward mitigation, the reduction of greenhouse gas emissions. However, mitigation is an illusion (Monastersky 2009: 1091, 1092).[1] Thus, all nations face the imperative of adapting to the inevitable impacts, negative and positive, of global climate change. Adaptation will require both new property laws and increased state regulation. Global climate change will have significant, although not fully understood (See United States Climate Change Program 2003), impacts on the physical form of much of the earth's surface, the uses to which land will be put, and the value of certain lands. The most dramatic impact is sea-level rise which could eliminate human settlements[2] and wetlands. For this reason, the global climate change literature initially focused on the impacts of vulnerable coastal land. This concern now extends to dry land.

Property in areas subject to more extreme droughts and flooding may become less valuable for irrigation farming or dry land farming respectively. A 2008 report estimates the yearly damage to California real estate and illustrates the range

---

1 The latest research suggests that we are reaching dangerous CO2 concentrations more quickly than previous estimations and that the recovery time from reductions, should they actually occur, may be as much as 1,000 years. See generally Braasch 2007 (providing more information on Global warming recovery times).

2 A 2009 study concludes that the rate of sea level rise along the Atlantic Coast in the twentieth century was three times higher than the previous 500 years and the magnitude of the rise is higher for the federal state of North Carolina, which has extensive recreation residential development along its low lying Outer Banks (Kemp et al. 2009). Another study, James G. Tutus, Environmental Research Letters, found that along the Atlantic Coast 60 per cent of the dry land within one meter of the current high tide line was developed.

of the landscape impacts of climate change common to vulnerable areas of the world. Excluding agriculture and forests, the Report estimates that global climate change damages will be US$0.2bn to 1.4bn for water and US$0.1bn to 2.5bn from fire (Roland-Holst and Kahrl 2008). The potential damage includes the risk of increased wildfires, sea level rise and the resulting coastal erosion and increased winter storm flood and wind damage. When forests and agricultural land damage are factored in, the yearly range is US$0.3bn to 4.3bn. New impacts continue to emerge. For example, public and private conservation reserves may become less functional as species migrate away from their intended warmer 'homes' (Ruhl 2008). In addition, land management is an increasingly important technique to reduce greenhouse gas emissions by sequestering carbon dioxide in forests and wetlands (Bosselman 2009).

The costs of these impacts will be unevenly distributed. Developing nations face the possibility that climate change's impacts on land will be greater than for developed nations, although there are a variety of factors such as crop substitution which must be considered before any firm conclusion is reached. A recent study estimates that Namibia's GNP could fall between 1 and 6 per cent due to reduced fishing and the reduction in livestock raising and subsistence farming (Johnson 2009). Some island nations may disappear and this raises the question of whether national sovereignty survives the disappearance of a nation's land base. Other countries may need to make adjustments in their borders. Austria, France, Italy and Switzerland have begun to discuss the alternation of their borders, demarked by glacier boundaries, as the ice retreats (Johnson 2009). There will also be winners among land holders. Land in the northern latitudes may be more valuable for agriculture as the growing season lengthens or population migration increases (Reilly 1996). In places as diverse as California and Namibia, prime irrigated agricultural land may become even more valuable as scarce water resources force the retirement of marginal lands.

Lawyers and climate policy students have addressed strategies to address sea level rise and have begun to speculate about the consequences of shifting and changing ecosystems. Less attention has been paid to the related question of whether global climate change will alter the law of property ownership and enjoyment. Land owners have always faced disasters and severe declines in the market value of land, but in general, disasters and value declines are assumed not to impact the legal ownership of property; ownership is an abstract interest in a thing and that interest remains as long as the thing remains. Both the common and civil law protect the owner's right to exclude (*jus prohibendi*) rather than the right to fixed value.[3] National constitutions and statutes only protect value when the government regulation triggers a duty to compensate (see Alterman 2010).

3 For example, Italian Civil Code Article 832. Civil law property rights are *im rem* rather than *in personam* rights (for example Taekema 2004). Thus, they run to the whole world and 'are effective with respect to anyone who interferes with the thing' (Kohler 2009: 229).

Thus, the assumption is that insurance or new government compensation schemes will be the primary vehicles to buffer property owners from these shocks.[4] Only in limited instances, almost exclusively dealing with water boundaries, does the existing law consider parcel loss from certain forces of nature to be a risk inherent in property ownership.

Property Rights: Risk versus Security

Property ownership always entails some risk that less than full, expected use and enjoyment of an entitlement will be recognized. Competing claims can displace title; even in the civil law countries with well-developed title registration systems, there is a risk of title failure. For example, Italian law can award title to the first person to register a sale even though there was a prior contract of sale and the first registrant knew of the contract.[5] Common law jurisdictions allow title to be lost through adverse possession. The doctrine is harder to justify in land registration systems, but many civil law systems allow prescription or *usucapation*.[6]

Global climate change will increase the risks of property ownership and land titles in at least two ways. First, it will expand the situations in which private boundaries change. Owners in vulnerable coastal and other water-related areas face increased risks that pre-climate change patterns of entitlements will no longer hold. A dramatic example of this is occurring in the Panhandle of the United States federal state of Alaska. When the area was covered with glaciers, there was a navigable channel between the capital, Juneau and a port ten miles away on the Inland Passage. Area glaciers are melting, and the channel and associated wetlands are silting in as the land rises due to the release of the pressure placed on it by the ice. The land that 'is emerging from the water to replace the lost wetlands [is] shifting property boundaries and causing people to argue about who owns the acreage and how it should be used' (Dean 2009).

Second, and perhaps more important, governments will be forced to use climate change science to require that property owners take more aggressive adaptation measures. Some measures, such as increased coastal armouring, will protect existing entitlements. Others, however, will try and move people out of harm's way or make it more expensive to build and live there. These could include the reduction of incentives to build (and to rebuild after a natural disaster) in vulnerable areas, to grow irrigated crops that demand large amounts of water in areas likely to face serious water shortages, or to continue the practice of constructing unsustainable, automobile dependent urban settlement patterns. This

---

4 Professor Daniel Farber is the leading United States exponent of victim compensation (Farber 2007).

5 Italian Civil Code. Section 1376.

6 Italian Civil Code. Section 1158.

in turn will conflict with one of the central aims of modern property: the creation of perpetual interests in land.[7]

Philosophical justifications for private, strong individual property rights abound (Penner 1997). The common objective of property law is an abstract,[8] uniform system of private control of resources. There is a widespread consensus that property law should serve three, interconnected overarching objectives in order to promote security of ownership: (1) the creation of individual, exclusive entitlements, (2) the creation of freely alienable rights which will support the use of markets to allocate scare resources, and (3) the demarcation of private entitlements secure from excessive inference by the Hobbesian sovereign.[9] The role of the state in defining and regulating the use of property is recognized, but most theorists assert that the incidents of the entitlement, including the public and private boundary, should remain *relatively* constant over time. These objectives have, of course, always been imperfectly realized due to the costs of enforcing property rights and the persistence of counter-objectives. Still, we think of property as a relatively static rather than dynamic relationship between a person and a thing. This principle was dramatically applied when a panel of The European Court of Human Rights found that Great Britain's relatively title holder adverse possession statute[10] was a violation of Article 1 of Protocol 1 of the European Convention on Human Rights which protects the peaceful enjoyment of property.[11] The Grand Chambre reversed and found that adverse possession was a reasonable state limitation on title security,[12] but the first decision is a reminder of the power of the idea of the subordination of state power to the protection of the security and stability of private entitlements.

Dynamism and Risk

There has always been a counter dynamic aspect to property rights. The Grand Chambre's decision in *Pye* recognized that seemingly secure property rights

7 Echeverria and Pidot, note that '[t]he goal of permanent land protection, while emotionally appealing, may turn out to be naive and misguided when the land itself is in flux' (2009: 10868, 10877).

8 The German Abstrakionprinzip, for example, recognizes the transfer of registered title even if the contract of sale is invalid (Kohler 2005: 230–231).

9 For the United States perspective see Ellickson 1993, Dagen and Heller 2001, Merrill 2000. In other countries, there is a greater recognition that property rights must be defined in relationship to the promotion of social welfare (Alterman 2010: 271–272).

10 Adverse possession is a doctrine found in common law, and to a lesser extent civil law systems. It allows a possessor who enters land owned by another to acquire title to the land by maintaining a certain standard of possession for the period prescribed by the applicable statute of limitations. At the end of the statutory period, the previous owner's possessory remedies are barred.

11 *Pye v. United Kingdom*, Judgment No. 44302/02, 13 November 2005.

12 *Pye v. United Kingdom*, Judgment No. 44302/02, 30 August 2007.

can be dislodged. However, the risks of disruption and change have always been treated as an exception to the stability principle. In general, common law property rights are more subject to change because the power of judges to create new rights and duties is greater than the power of civil law judges. Property owners have historically faced two types of risks to the stability. The first is a change in the law which decreases exclusivity. The second is that states may exercise their sovereign power to limit the use of property. An example of the first category occurred when the Supreme Court of the federal state of Ohio rejected a trespass suit by owners of land against an oil refinery.[13] Underground injections caused a plume of contaminated substances to migrate beneath the plaintiffs' land, but no loss of use, decrease in value or even a public health risk was proved. The owners demanded compensation nonetheless because trespass protects the expectation of exclusive possession, in contrast to nuisance law, and thus no actual damages need be shown and Anglo-American common law recognizes that one owns both upwards and downwards of the surface of the earth. But, the Court found the principle 'inapplicable'. 'Just as a property owner must accept some limitations on the ownership rights above the surface, we find that there are also limitations on property owners' subsurface rights'. Thus, plaintiff's right to exclude is limited to invasions that actually interfere with reasonable and foreseeable surface use.

The second category includes the whole range of land-use regulation found in all developed countries. States possess the inherent power to regulate the use of land. State power is constrained by constitutions and legislation which provide remedies for landowners who have suffered the substantial loss of value from regulation. The level of protection varies from country to country (Alterman 2010), but it is widely accepted that the state owes a duty of compensation in situations where an individual land owner has suffered a disproportionate loss of the value of his or her property. The loss of the ability to obtain the maximum potential value of a piece of land is seldom an occasion for compensation.

Usufructory Rights: Water Boundaries, Water Rights and Servitudes as Case Studies in the Inherent Risks of Ownership

Water Boundaries

Water boundaries can never be permanently fixed for two reasons: rivers, lakes and the sea are constantly changing adjoining land, and in all countries, the state owns the submerged lands of the sea and large rivers and lakes up to some boundary standard such as the mean high tide line or the high water mark.[14] Thus, water boundaries on lakes, rivers and oceans are dynamic rather than static. Property

13 *Chance v. BP Chemicals*, 670, N.E.2d 985 (Ohio 1996).
14 For example, *Brickell v. Trammel*, 82 So. 221 (Fla. 1919).

owners may lose and gain soil due to erosion and reliction or the deposit of soil. The common and civil law classifies these processes as accretion or avulsion.[15] A riparian land owner is entitled to the gradual deposit of soil that extends his original boundary.[16] If water retreats, the land owner gains the newly exposed soil under the doctrine of reliction. Conversely, the owner may lose a portion of his original land if the stream course migrates landward. However, in rapid shifts, avulsion, boundaries do not shift.[17]

Ironically, efforts to deal with global climate change may require the state to create stable water boundaries that push private property boundaries landward. Florida's legislation to do this and the subsequent takings challenge are instructive. The state enacted legislation which allows the state to freeze the public-private water boundary – the mean high tide line – and to establish an erosion control boundary (ECL), and declared that all artificial beach nourishment seaward of the ECL remains the property of the state. Littoral owners have a right of access along with the general public but no right to future accretions. Property owners challenged the ECL as an unconstitutional taking of private property. Although Florida had held that littoral rights such as view cannot be taken without compensation,[18] the state Supreme Court classified view as a present right and the right to receive accretion as a contingent future one and ancillary to the right of access. This made is easier to conclude that the state had struck a reasonable balance between the public interest in preventing beach loss, the reduction of storm damage and the conservation of the shoreline and the property owner's interest beach access.[19] It also noted that hurricanes are avulsive events so the property boundary remains in place but the property owner has a reasonable period of time to reclaim the lost land. Thus, the legislation simply allowed the state, as public trust owners of land seaward of the mean high tide line, to do what the common law allows them to do. The United States Supreme Court affirmed because Florida Supreme Court decision was consistent with 'background principles of state property law … It did not abolish Members' right to future accretions, but merely held that the

15 The Roman, civil and common law origins of the doctrine are briefly traced in Angell 1983.

16 The rationales are (1) *de minimus non curat lex*, (2) fairness dictates the right to accretion as compensation for erosion, (3) all property should have an owner and for convenience the riparian is the owner of accretion, and (4) the necessity to preserve water access. *Board of Trustees of the International Improvement Trust Fund v. Medeira Beach Nominee*, 272 So.2d 209 (Florida Court of Appeals 1973).

17 After an exhaustive examination of the origins of the avulsion-accretion distinction, Professor Joseph Sax concludes that it should be abolished, except in limited cases, and land titles should follow moving water boundaries to retain a riparian's 'water adjacency' (Sax 2010b: 305, 353).

18 For example *Lee County v. Kiesel*, 705 So.2d 1013 (Florida District Court of Appeals 1998).

19 *Walton County v. Stop the Beach Nourishment, Inc.*, 998 So.2d1102 (Fla. 2008).

right was not implicated by the beach restoration project, because the doctrine of avulsion applied.'[20]

Water Rights

The law of water rights is another example of a property rights system which expressly incorporates the risk of non-enjoyment of the resource as well as of physical alteration directly into the legal right. Both the right to use water and the delineation of water boundaries carry with them the risk that the physical nature of the right, and thus its use and enjoyment, will not remain constant over time. All water right holders face the risk that the entitlement may be curtailed due to the vagaries of climate or the demands of competing users and uses (Tarlock 2010). For this reason, water rights have always been incomplete rather than complete property rights[21] because they are correlative; one person's use is subject to the correlative rights of others. In addition, the fact that water is unique and necessary for survival gives the state great discretion to establish the ground rules for the acquisition and exercise of water rights so as to incorporate the interests of other users and the public generally in private rights.

The fugitive nature of water makes water physically and legally different from land. Most legal systems assert that the public owns the waters of a territory.[22] State ownership is a fiction for the assertion of the power to regulate all aspects of use and enjoyment. Private rights are not excluded but they are usufructory (Caponera 2003: 83, 84) rather than full, exclusive ownership. This said, however, water law is an example of a property regime which forces adaptive behaviour.

There are two major systems of United States water rights, and both illustrate the inherent risks in the use of water to which users must constantly adapt. The common law of riparian rights is a land ownership-based system which, modified by legislation, prevails in the eastern, humid United States. All land owners along a stream have an equal right to use the water, but the right is defined by general and open-ended reasonableness standards. The right can be firmed up only through judicial action. The result is that the right to use water is highly uncertain because courts can, in theory, displace prior users to make room for subsequent ones or changed conditions such as global climate change.

Riparian law has remained vague because there has been sufficient water to meet all competing demands. However, the incoherent nature of riparian rights discourages serious risk analysis. Many states have replaced the common law with

---

20 *Stop the Beach Renourishment, Inc. v. Florida Department of Environmental Protection*, 130 S.Ct. 2592, 2601 (2010).

21 In re Water Use Permit Applications, Petitions for Interim Instream Flows Standard Amendments, and Petitions for Water Reservation from the Waihole Ditch, 94 Haw. 97, 9 P.3rd 409 (2000).

22 For a discussion of Italian water law see Lugaresi 2000, Islamic law is discussed in Caponera 2003.

permit systems. Regulated riparianism brings some certainty to the system because in practice it introduces a greater measure of stability through the protection of prior users, but risk remains because the duration of state water licenses is limited so there is no firm expectation that they will be renewed.

Prior appropriation is the opposite system and is, in theory, a complete global climate change risk allocation scheme. According to the catechism, prior appropriation allocates water in times of shortage by strict enforcement priority schedules which provide fair notice to junior users of their potential risks. However, prior enforcement is generally confined to small systems. The western states have worked hard to make sure that there are few calls for enforcement, and thus the law has never been used as a risk allocation system. There are major political, institutional and legal barriers to using the law to distribute the extreme risks of global climate change among large classes of water users The thrust of federal and state water policy from the conservation era[23] to the 1980s has been to minimize the risks of shortages by constructing large carryover storage facilities. Thus, reservoirs and groundwater basins probably will be subjected to only the mildest form of rationing during droughts. In addition, formal and informal mechanisms share the burdens of shortages by pro rata rather than pro tanto delivery reductions. Given the expectations that there will be limited enforcement of priorities, junior water right holders will resist the consequences of hard prior appropriation, and states will try to avoid strict enforcement of the system when the results are inefficient.[24]

23 In the United States, the conservation era refers to the period between 1890–920 when the federal government began to regulate the use of public land and to promote the idea of multiple purpose resource use. This led to the construction of large reservoirs for flood control, hydroelectric power generation and the storage of water for municipal and irrigation use. The leading history is Hays 1959.

24 The state of Idaho's efforts to resist a call in the Snake River plain are a case in point. Starting in 1993, senior appropriators have made calls on junior pumpers. The initial calls have been averted but creative regulations and a court acutely aware of the limits of prior appropriation, but the possibility of future calls is real. In 2005 two trout farms in the Magic Valley in the south central part of the state made a call and rejected an initial offer of 45,000 acre feet of replacement water. The Idaho Department of Water Resources eventually threatened to shut pumps for 33,000 acres and several towns and industries in the Valley. To try and avoid this drastic remedy, the state adopted Rules for the Conjunctive Management of Surface and Ground Water Resources. In brief, the rules allow the Director of the Idaho Department of Water Resources to apply a combination of the traditional futile call doctrine and resurrected the doctrine, ironically developed in the Snake River Plain, that a senior's use mean of diversion must be reasonable before a call will be honoured. *Schodde v. Twin Falls Land and Water Co.*, 224 U.S. 107 (1912) (senior water right holder not entitled to current necessary to run water wheel which lifted water from a canyon to the rim). A district court held that the Rules violated the constitutional right to divert because they do not permit the timely administration of water rights and failed to include a presumption that any junior withdrawal in times of shortage is a per se

Servitudes

Servitudes are lesser interests compared to ownership. A servitude allows one to use or control the use and enjoyment of the land of another. They are thus usufructory rights, and as such, have always been more fragile compared to ownership rights. The right to use is limited in scope.[25] Ownership cannot be abandoned but global climate change introduces a new risk for servitude holders. In the past 40 years, United States conservation organizations have acquired millions acres of development rights in order to preserve land in open space and to provide habitat for flora and fauna. Studies now suggest that as species migrate the purpose for which many easements were acquired can no longer be served. Therefore, owners of the underlying ownership right could claim the land free of the servitude. Many easements have clauses that prevent termination in the event of changed conditions on the land. One proposal is to characterize conservation easements as arks not parks. Instead of clauses that virtually prohibit termination in the case of changed circumstances, the easement could be terminated only if global climate change has caused the species originally intended to be protected to migrate (Olmstead 2009: 10842, 10846).

The Importance of the Incorporation of Moral Hazard into the Law of Property Rights

The Need for Moral Hazard Analysis

The prospect of sea level rise from thermal expansion, glacier melt and the loss of polar ice sheets is one of the most serious adverse impacts of global climate change. Vulnerable areas include island nations, the North Sea, the Atlantic Coast of the United States, and coastal population concentrations in Bangladesh,

---

interference with senior surface rights, but the Supreme Court reversed. *American Falls Reservoir Dist. v. Idaho Dept. of Water Resources*, 143 Idaho 862, 154 P.3d 433 (2007). The Court first observed that '[w]hile the Constitution, statutes and case law in Idaho set forth principles of the prior appropriation doctrine, these principles are more easily stated than applied. These principles become especially more difficult, and harsh, in their application in times of drought.' The need for a presumption of interference was rejected because Director of the Department of Water Resources, the state engineer of old, needed the discretion to decide when to honour a call and thus the rules contained sufficient standards and did not constitute

25 United States water access and use easements have long been limited to the use contemplated in the original grant. For example, *Stefanoni v. Duncan*, 282 Conn. 696, 923 A.2d 737 (2007) (water access easement does not include right to maintain a walkway and dock which would burden servient owner's littoral rights); *Brannon v. Boldt*, 958 So.2d 367, 374 (Fla.App. 2 Dist. 2007) (Back lot access easement to water does not include lingering on beach 'for extended period of time to view the water, fireworks, or the sunset'.)

Vietnam and China (Nicholls et al. 2007). Coastal areas have no choice but to adapt. The basic choice is between armouring the shore or retreat. Armouring will be the initial choice of almost countries, but the costs will be high as illustrated by a new analysis on the impacts of sea level rise on the United States Atlantic Coast:

> Our results suggest that the majority of low-lying lands along the US Atlantic coast will become populated if business-as-usual development continues. Maintaining this development as sea level rises would require increasingly ambitious shore protection. The US experience protecting populated areas below sea level from flooding is mostly limited to metropolitan New Orleans. Sea level rise could leave communities similarly vulnerable throughout the US Atlantic coast. The resulting shore protection could imperil a key environmental objective in the United States: the preservation of tidal wetlands. In the 1970s, the United States collectively decided to stop creating new coastal communities by filling marshes and swamps, and enacted other policies to preserve tidal wetlands along the Atlantic coast. But these ecosystems may not be sustained if sea level accelerates. At the current rate of sea level rise, most tidal wetlands are able to keep pace through sedimentation and peat formation; but their ability to keep pace with a rate greater than 5–10 mm yr–1 is doubtful. (Titus et al. 2009)

To survive, these ecosystems would have to migrate inland. With only 9 per cent of the lowest land set aside for conservation, a large-scale migration would require either a halt to construction in most coastal floodplains or an eventual abandonment of many developed areas. But current policies promote the opposite. Many serious environmental students counsel modified forms of retreat or the adoption of the Dutch model which includes a full range of adaptation strategies including selective flooding.[26] A long-time resident of New Orleans and leading environmental law scholar, Professor Oliver Houck, envisions a future for southern Louisiana built around less retardation of natural water flow and withdrawal of human population from high risk areas (Houck 2006). However, as he observes, 'it will be nearly insane in a region that equates planning with socialism and has always looked on the Corps for a bigger fix' (Houck 2006: 61). Actions taken by officials in Galveston, Texas in the aftermath of Hurricane Ike to rebuild an extremely vulnerable urban area illustrate the constraints that the political process places on efforts to create incentives to minimize the damages caused by extreme natural events in advance of their occurrence.[27]

---

26 See European Union 2009; Olsthroon, et al. 2006 concludes that retreat may be the most cost effective strategy for some areas of Holland.

27 See Been for details how efforts to protect South Carolina's fragile beaches were rolled back after *Lucas* contrary to the theory that takings law promotes the adoption of efficient regulatory programs (2004: 221).

Legal Barriers to the Incorporation of Moral Hazard into Public and Private Decisions

The law frustrates serious adaptation in two ways. First, legislatures in developed countries have created the expectation that victims of natural disasters should be restored as close to status quo as possible through public funds above voluntary insurance payout levels (see Faure and Hartlief 2006). Second, the constitutional, treaty and legislative protection of property can both encourage land-use practices which increase the costs of a natural catastrophe and permit challenges to regulatory measures designed to prevent and mitigate these costs. Neither disaster relief nor the constitutional protection of property gives adequate weight to the moral hazard problem in property rights law.

The Concept of Moral Hazard

Moral hazard is a concept first developed by the insurance industry to avoid excessive liability in situations where the insured has the capacity to take preventative action but has little or no incentive to do so. For example, a home owner who obtains fire insurance has some incentive to engage in behaviour that may increase the risk of a fire, and fewer incentives to refrain from activities that may cause a fire or to take preventative measures. To minimize the risks of this moral hazard, insurance companies have taken steps, such as deductibles, to 'incentivize' property owners to be more careful. Today, moral hazard applies to any risky behaviour which could be avoided but is not because of the expectation that such behaviour will not be penalized. In fact, the behaviour is often encouraged by laws which immunize the actor from the full or partial responsibility for avoiding the costs and which shift the costs to the public funds. Thus, a moral hazard can be defined as socially undesirable or inefficient behaviour which is encouraged either by the expectation that it will not be punished or, more perverse yet, that it will be rewarded (Byrne 2003: 679, 687).

The Need to Incorporate Moral Hazard into Public Policy

Efforts to design more environmentally sustainable, resilient landscapes that have the capacity to adapt to global climate change seek to do no less than to impose the economic doctrine of moral hazard on public and private land-use decisions. For example, a noted American geographer, the late Gilbert White,[28] long argued that it is economically irrational to encourage people to locate in the face of danger, such as floodplain or a vulnerable hurricane zone, and who then demand to be

28 The late Gilbert White pioneered the idea that nations must try and keep human settlements out of floodplains protected by dams and levees because when floods inevitably occur, the amount of damage will be substantial because of the human settlement patterns that the structures encourage (White 1942).

compensated when damage occurs. The legal and political challenges to reverse the incentives to constructing a landscape that can adapt to natural disasters, especially those disasters that may be enhanced by global climate change, are formidable. The reason is that land-use planning law often creates too many incentives from land owners to assume bad, predictable risks and very few incentives to avoid the consequences of the risk.

Takings Law: An Example of Moral Hazard Disrespected

United States takings jurisprudence is a classic study in the failure to incorporate moral hazard into the law. The incentives for property to assume risks starts with the 5th Amendment to the US-American federal Constitution and continues through the well-justified expectation that the federal government will compensate a wide range of natural disasters. In between is the long history of the construction of flood control projects and a federal flood insurance program that still encourages building in high risk areas (Wish 2009). The problem is not with the basic idea of helping victims of natural disasters, but with our inability to distinguish between deserving victims and subsidized risk takers. The basic idea of compensation law is the promotion of fairness. The law is designed to compensate victims of regulation who have suffered substantial and *unanticipated* losses in the value of their property, which are disproportionate in comparison to those suffered by similarly situated landowners. These candidates for compensation are on the whole individual property owners who have been required to confer public benefits, or who are denied equal protection of the law by regulation which comes too late to address the problem that triggered the need for regulation (Sax 2010a: 455).

Moral hazard is a different situation, but the Supreme Court has been reluctant to incorporate moral hazard expressly in its jurisprudence, although the concept is in fact embedded in its jurisprudence. The Court has expressly said that an important element of fairness is whether the victim has legitimate invested backed-expectations that a compensable property will be recognized. The explicit incorporation of invested-backed expectations in 1978 indirectly introduced the idea of moral hazard into takings law.[29] The question then becomes: can the state devalue property and avoid compensation because the property owner should have expected the regulation and not invested in the project or activity. The Court both limited and expanded the expectations analysis in *Lucas v. South Carolina Coastal Council*.[30] The issue was whether a state beachfront set back statute could be applied to limit development on a barrier island. The Court held that the statute could not be applied to block the construction of a house close to the ocean, because

29 *Penn Central Transportation Co. v. New York City*, 438 U.S. 404 (1978). The Supreme Court identified '[t]he economic impact of the regulation on the claimant, and, particularly, the extent to which the investment has interfered with investment backed expectations' are the major relevant factors is deciding whether compensation is due.'

30 505 U.S. 1003 (1992).

the complete elimination of the value of property for development is a categorical taking.[31] The case is actually a simple equal protection case, because the island was almost fully developed with unstable beach front stilt houses before the state decided to limit coastal development. The majority opinion did, in dicta, recognize that there is a limited class of common law background title limitations[32] which can be raised as defences to compensation. The duty not to use one's property to cause a nuisance is one such limitation, and that was the basis of South Carolina's defence of its law, but the majority opinion refused to accept the state's rationale that the statute prevented the destruction of other property during hurricanes.

Most of the critical environmental commentary has focused on the Court's hostility to the idea of all environmental regulation which is unaccompanied with full compensation, but the Court's limited articulation of the moral hazard dimensions of 5th Amendment jurisprudence may be a larger, long term problem. The resistance to moral hazard analysis continued with the Supreme Court's dismissal of the argument that the purchaser of highly regulated property assumes the risk of development denial with the quip that '[t]he state may not put so potent Hobbesian stick into the Lockean bundle'.[33] Locke himself might be surprised that his labour theory now incorporates the Roman law right of *ius abutendi*, the right to destroy property (Strahilevitz 2005: 781). This analysis and the *Lucas* plurality's rejection of the state's damage prevention argument illustrate that the modern notion of property remains rooted in the notion of exclusive dominion subject only to the duty not to cause a nuisance. Recent litigation in Texas illustrates the legacy of this thinking. Public beach access has been a major issue in the United States, and states have used different strategies to expand public access of the dry sand area. Texas law recognizes public prescriptive rights to use private beaches[34] up to the vegetation lines. A court interpreted the law to mean that public easements move with changes in the vegetation line.[35] But, a recent case allowed a constitutional challenge to the rolling easement doctrine by a property owner whose property migrated landward after Hurricane Rita. Over a strong dissent, an extremely conservative federal court of appeals judge (blocked for the Supreme Court during the Bush II administration) characterized Texas law as ambiguous

31 Prior to *Lucas* only physical occupations were categorical takings. *Loretto v. Telprompter*, 458 U.S. 419 (1982). (Unconsented cable television cable on exterior of building.)

32 Blumm and Ritchie survey the post-Lucas cases and find that the background limitation defence is growing, many of the examples of the defence involve resources, such as water, fishing quotas and public land mining, where the expectation of exclusive dominion has always been lower than it has for land (2005: 321).

33 *Palazzolo v. Rhode Island*, 533 U.S. 606, 627 (2001). In her concurrence, Justice O'Connor argued that the property owner's knowledge the extent of regulation was an element to be considered in determining the compensation, if any, to which he was entitled under the interference with investment backed-expectations standard.

34 Texas Natural Resources Code s 61.011.

35 *Arrington v. Texas General Land Office*, 38 S.W.3d 764 (Tex.App. 2001).

and uncertain, held that a landowner could not bring a 5th Amendment taking claim because the issue was not ripe, but that the state's attempted enforcement of the easement was arguably a Fourth Amendment seizure.[36] Observers of European law detect a convergence with United States regulatory jurisprudence as a result of The European court of Human Rights (Jacobs 2009: 3).[37] In addition, the same insulation from foreseeable regulation, including climate change adaptation, can be found in international investment law. International Investment Agreements may contain clauses that require compensation from non- discriminatory and 'foreseeable' changes in the law (Johnson 2009: 11153).[38]

Land-use Controls Law Does Not Deter Moral Hazards

Land-use regulation will be an important component of global climate change adaptation. First, governments are being urged to identify high risk areas, such as areas vulnerable to flooding, and to deflect or limit building in these areas. Second, the design standards for permitted buildings will become more stringent. However, land-use controls, especially zoning, are ill-suited to this objective. Zoning is a product of the Progressive Era. It was born in the major metropolitan areas and reflected the untested idea that a well-planned built environment would make people's lives better,[39] but it did so in ways that placed minimal limitations on location choice. Specifically, it was built on two ideas, both of which are ultimately grounded in the common law of nuisance: the protection of the single family neighbourhood from inferior uses (and people), and the protection of superior real estate from inferior parasitic uses. Much urban and suburban zoning still revolves around these issues, although the underlying progressive rationale has long since vanished. In fact, the whole idea of the need for rigid use segregation has long been discredited. Zoning has evolved, and courts have long held that the common law of nuisance is not the limit of the police power. However, the fact remains that zoning in the United States was never developed as a tool to regulate non-urban landscapes on a comprehensive scale. Zoning and subdivision controls have evolved considerably, but cities still are limited to tinkering with the pace and density of individual developments and to 'taxing' them within the limits of

---

36 *Severance v. Patterson*, 566 F.3d 490 (5th Cir. 2009). Procedurally, the federal court certified to the Texas Supreme Court for resolution. See generally Hofrichter 2009.

37 The contrary theory was articulated by the distinguished comparative planning expert Rachelle Alterman (1997).

38 The author speculates that climate change ,might be a basis to invoke the customary national security exception to treaty obligations.

39 Fairfield (1993) demonstrates that zoning represented the partial triumph of the progressive ideal that expert government intervention is necessary to temper the operation of the market and that rational, scientifically-based planning and intervention can improve people's lives.

Nollan-Dolan[40] legal exactions. But most regulatory decisions are still relatively small-scale, individual adjustments. The result can be compared to fitting the pieces of a jig-saw puzzle together with no expectation that a clear picture will ever emerge. Put differently, zoning is still heavily influenced by the common law of nuisance. Regulation can prevent the undesirable side-effects of certain uses much better than it can create a sustainable built urban environment, or conserve a rural one. For example, the current landscape, which is the product of a long chain of federal, state and local policies, has been much criticized as unsustainable. But, continued sprawl is strongly defended both as the best way to promote low and moderate income housing, and even as a rational post 9/11 strategy to mitigate the damages from a terrorist attack (Ziegler 2006).

The Need to Incorporate Moral Hazard into Land-use Planning

To adapt to global climate change, governments will have to prepare land-use plans designed to mitigate damage from CGG-caused events such as intense storms, sea level rise, and drought, *and* require that all development in this area be consistent with the plan. For decades, planners have argued that regulation should be subordinated to planning. The argument has borne some fruit in legislation and judicial decisions, and there are many large-scale planned communities being built in growing areas. However, many courts and statutes treat plans as advisory or adopt liberal standards of consistency. Perhaps more important than consistency, global climate change adaptation land-use plans must specify the risks that individual property owners must assume in order to minimize the moral hazard. It is too early to speculate how global climate change might change the nature of private entitlements. But, it seems clear that societies will have less tolerance for subsidizing moral hazard risk taking. It seems equally clear that societies will have to find a new balance between offering compensation to victims of natural disaster, and asking them to bear an increased cost in cases where the victim could have done more to avoid the losses.

One can see this legacy in the United States National Flood Insurance Program (Kunreuther and Roth 1998; Mileti 1999). In 1968, Congress enacted legislation to induce property owners to reduce flood damages by either building flood-resistance structures or not building in floodplains. However, the vehicle chosen was federally subsidized insurance for those living in mapped floodplains.[41] A recent coalition of investors and insurances called for more accurate, climate change adjusted, floodplain maps, for more aggressive local land-use policies that bar new flood development in high risk areas, for rebuilding after a flood, and for storm resistant

40 Two United States Supreme Court cases, *Nollan v. California Coastal Commission*, 483 U.S. 825 (1987) and *Dolan v. City of Tigard*, 512 U.S. 374 (1944), require a substantial nexus between the social cost of development and the fees imposed by a local government to recover part of these costs.

41 The program has a $19bn deficit.

construction standards, but as of 2011 Congress has not acted (Heinz Center 2010). But, in the future, private insurance will play an increasingly important role in forcing property owners to avoid moral hazards.[42] Insurance rates will reflect the risks inherent in building in floodplains and other vulnerable areas. However, the United States has often subsidized risk taking. After several major hurricanes struck the federal state of Florida in 2004–2005, private insurance companies withdrew from issuing water damage hazard insurance. The state stepped with its own insurance program which now has a $500bn risk 'portfolio' (Lehmann 2009). Initially, the rates were discounted and frozen by the state legislature, but in 2009, the legislature allowed the agency to raise its rates 10 per cent a year until it is financially sound. After the legislation, the state proposed a 7.5 per cent increase. Experts, however, predict that the agency might have to double premiums of high risk homes before it will have the reserves to meet the damages that future, perhaps global climate change enhanced, hurricanes will cause.

New Legal Protections for Property

The primary focus of this paper is how global climate change may change the physical or legal structure of property rights or increase the risk of enjoyment, but global climate change may also increase the legal protection of property at risk from global climate change. In the United States, environmental groups have been frustrated with the inability of Congress to pass legislation to mitigate and adapt to global climate change; they have increasingly turned to the common law for relief, although the main purpose of these actions may be to increase pressure on legislatures to act and pre-empt common law actions.[43] In the aftermath of Hurricane Katrina in 2005, a group of property owners along the Mississippi Gulf Coast brought a state common law nuisance suit against several oil and chemical companies alleging that the operation of these activities 'caused the emission of greenhouse gases that contributed to global warming, *viz.*, caused a rise in sea levels and added to the ferocity of Hurricane Katrina'.[44] The case was brought in federal district court and the plaintiffs had to establish standing under both state and federal law and to convince the court that the action should not be dismissed, as

42 In January 2010, the United States Securities and Exchange Commission announced that it will issue an interpretive guidance on the disclosure of climate change impacts and risks in securities disclosure documents. Commission Evidence Regarding Disclosure Related to Climate Change, 17 Code of Federal Register Parts 211, 231 and 241 (2010).

43 Two lawyers in a leading Chicago law firm have observed that the common law actions discussed below 'made real the prospect of tort liability to private parties seeking to recover for alleged damages of global warming'. Sigel and Cassel 2009:28.

44 *Comer v. Murphy Oil USA*, 585 F.3rd 855 (5th Cir. 2009). A federal district court in California had earlier dismissed a nuisance suit by an Alaskan village, which is flooding, against various oil companies because it presented only a political question. *Kivalina v. Exxon Mobile Corp.*, 2009 U.S. lexis 99563 (N.D. Cal. 2009).

the trial court held, because it was a non-justifiable political question. The Court of Appeals found standing under both state[45] and federal law. The defendants argued that Article III requires that injury be fairly traceable to the defendant's actions, and thus plaintiffs had to make a full causal showing to support their allegation. However, the Court held that common law tort and property claims seldom raise standing problems, and that the fairly traceable standard does not require proof of either cause-in-fact or proximate cause. Rather, the allegations are presumed to be true and were sufficient to support standing because it was plausible that air emissions 'causes or contributes' to the injury.[46] However, the case was dismissed on en banc review.[47]

Conclusion

Global climate change requires that we revisit the very foundation of property. Anglo-American property law is grounded in the thinking of Hume and Bentham, although there is a great deal of natural law noise. Hume argued that we recognize property because otherwise no one would work and society would be worse off. Bentham took Hume's utilitarian theory of property as a mental state and reduced this to his famous phrase 'Property is nothing but the basis of expectation.' Expectation is a function of law and thus can change over time. Thus, property is positive, political not pre-political as Locke posited. Even the strongest proponent of firm property rights, Professor Richard Epstein, recognized that any natural rights theory, such as Locke's labour theory, is circular because it assumes a right to possession of self and further 'property may look to be an individualistic institution, but the very nature and definition of the right seems to require some collective social institution to be at its base'. There is precedent for the recognition that extraordinary and long lasting conditions such as global climate change impact the expectations of title security and limited state regulation. Courts have

45 Mississippi does not limit the judicial power to cases or controversies and requires only that plaintiffs establish that they experienced an adverse effect from the conduct of the defendant.

46 The Court relied on the Supreme Court's precedent setting decision *in Massachusetts v. EPA*, 549 U.S. (2007), which accepted as plausible the allegation of a group of coastal states that the federal Environmental protection Agency's failure to regulate greenhouse emissions in automobiles and other sources contributed to twentieth century sea level rise.

47 *Comer v. Murphy Oil, USA*, 585 F.3d 855 (5th Cir. 2009) reh'g en banc granted, 598 F.3d 208 (5th Cir. 2010), and panel opinion vacated and trial court dismissal of action reinstated because seven panel member recusals deprived Court of en banc quorum. 607 F.3d 1049 (5th Cir. 2010). A 2011 Supreme Court decision, *American Electric Power Co. v. Connecticut*, 564 U.S. ___ (2011), signals that climate change tort suits will face substantial barriers. The case held that several northeastern states could not bring a federal common action against a public utility for CO2 emissions, although the opinion recognized that plaintiffs could proceed under state tort law.

increasingly refused to recognize abstract property rights when there was no reasonable expectation of the interest and there were limited demoralization costs to non-recognition. It is a major but logical step to adjust expectations to the risk of boundary loss, increased state regulation, and greater owner responsibility for mitigating the expected impacts of global climate change.[48]

References

Alterman, R. 1997. The challenge of farm land preservation: lessons from a six-nation comparison. *Journal of the American Planning Association*, 63(2), 220–243.

Alterman, R. 2010. *Takings International: A Comparative Perspective on Land Use Regulations and Compensation Rights*. American Bar Association, Section of State and Local Government Law.

Angell, J.K. 1983. *A Treatise on the Right of Property in Tide Waters, and in the Soil and Shores Thereof*. Littleton, CO: F.B. Rothman.

Been, V. 2004. Lucas v. The Green Machine: Using the takings clause to promote more efficient regulation?, in G. Korngold and A.P. Morriss (eds), *Property Stories*. New York: Foundation Press. 221–258.

Blumm, M.C. and Ritchie, L. 2005. Lucas's unlikely legacy: The rise of background principles as categorical takings defenses. *Harvard Environmental Law Review*, 29, 321–368.

Bosselman, F. 2009. Planning for a bull market for wetlands. *Planning and Environmental Law*, 61(4), 3–14.

Braasch, G. and McKibben, B. 2007. *Earth under Fire: How Global Warming is Changing the World*. Berkeley: University of California Press.

Byrne J.P. 2003. Property and the environment: Thoughts on an evolving relationship. *Harvard Journal of Law and Public Policy*, 28 (2), 679–690.

Caponera, D.A. 2003. *National and International Water Law and Administration: Selected Writings*. The Hague, New York: Kluwer Law International.

Dagen, H. and Heller, M.A. 2001. The liberal commons. *Yale Law Review*, 110, 549–623

48 My colleague Holly D. Doremus has identified some of the relevant factors that justify regulatory adjustments to changed conditions without compensation and serve as a starting point to analyse the role that global climate change may play in changing the law of regulatory takings. They are (1) justification for the change, (2), the foreseeability and ability of property owners to adapt to change, (3) the abruptness of the change, and (4) the general nature of the new regulation (Doremus 2003). The germ of this analysis can be found in the Rhode Island trial court's decision on remand from Palazzolo discussed in Note 33. The trial court found that proposed fill would be a public nuisance (Ruhl 2008: 753, 776).

Dean, C. 2009. Higher seas? As these glaciers melt, land rises. *The New York Times*, May 19, A1, col. 2.

Doremus, H.D. 2003. Takings and transitions. *Journal of Land Use and Environmental Law*, 19(1), 1–46.

Echeverria, J. and Pidot, J. 2009. Drawing the line: Striking a balance between regulating and paying to protect land. *Environmental Law Reporter*, 39(9), 10868–0881.

Ellickson, R. 1993. Property in land. *Yale Law Review*, 102, 1315–400.

European Union 2009. *The Economics of Climate Change in EU Coastal Areas*. [Online]. Available at: http://ec.europa.eu/maritimeaffairs/climate_change/netherlands_en.pdf [accessed 22 September 2011].

Fairfield, J.D. 1993. *The Mysteries of the Great City: The Politics of Urban Design, 1877–1937*. Columbus: Ohio State University Press.

Farber, D.A. 2007. Adapting to climate change: Who should pay? *Journal of Land Use and Environmental Law*, 23(1), 1–38.

Faure, M., and Hartlief, T. 2006. *Financial Compensation for Victims of Catastrophes: A Comparative Legal Approach*. Wien, New York: Springer.

Harvey M.J. 2009. An alternative perspective on United States–European property rights and land use planning: differences without any substance. *Planning and Environmental Law*, 61(3), 3–12.

Hays, S.P. 1999. *Conservation and the Gospel of Efficiency: The Progressive Conservation Movement, 1890–1920*. Pittsburgh, PA: University of Pittsburgh Press.

Heinz Center 2010. Resilient Coasts: A Blue Print for Action. [Online]. Available at: www.ceres.org [accessed 22 September 2011].

Hofrichter, M. 2009. Texas's open beaches act: proposed reforms due to coastal erosion. *Environmental and Energy Law and Policy Journal*, 4(1), 147–156.

Houck, O. 2006. Can We Save New Orleans? *Tulane Environmental Law Journal*, 19(1), 1–68

Johnson, L. 2009. International investment agreements and climate change: The potential for investor-state conflicts and possible strategies for minimizing it. *The Environmental Reporter*, 39, 11147–11160.

Kemp, A. et al. 2009. The timing and magnitude of recent accelerated sea-level rise (North Carolina USA). *Geology*, 37(11), 1035–1038.

Kohler, J. 2005. Property Law (Sachenrecht), in M. Reimann, and J. Zekoll (eds), *Introduction to German Law*. The Hague, Frederick, MD, USA: Kluwer Law International. 227–250.

Kunreuther, H., and Roth, R.J. 1998. *Paying the Price: The Status and Role of Insurance against Natural Disasters in the United States*. Washington, DC: Joseph Henry Press.

Lehmann, E. 2009. *Florida Tries to Escape its Exposure to Storm Damage.* [Online]. Available at: www.climatewire.net [accessed 22 September 2011].

Lugaresi, N. 2000. *Rethinking water law: the Italian case for a water code 106.* Universita Degli Studi Trento. [Online]. Available at: http://eprints.biblio.unitn.it [accessed 22 September 2011].

Merrill, T. 2000. The landscape of constitutional property. *Virginia Law Review*, 86(5), 885–999

Mileti, D.S. 1999. *Disasters by Design: A Reassessment of Natural Hazards in the United States*. Washington, DC: Joseph Henry Press.

Monastersky, R. 2009. A burden beyond bearing. *Nature*, 458, 1091–1094.

Nicholls, R.J. et al. 2007. Coastal systems and low lying areas, in M.L. Parry et al. (eds), *Climate Change 2007 – Impacts, Adaptation and Vulnerability: Working Group II Contribution*. New York: Cambridge University Press.

Olmstead, J. 2009. Perpetuity, latent ancillary rights, and carbon offsets in global warming era conservation easements. *Environmental Law Reporter*, 39, 10842–10850.

Olsthroon, X., Van der Werff, P., Bouwer, L.M. and Huiteman, D. 2005. Neo-Atlantis: Dutch responses to five meter sea level rise. [Online]. Available at www.fnu.zmaw.de/fileadmin/fnu-files/publication/working-papers/waisholland wp.pdf [accessed 22 September 2011].

Penner, J.E. 1997. *The Idea of Property in Law.* Oxford, New York: Oxford University Press.

Reilly, J. 1996. Climate change, global agriculture and regional vulnerability, in *Global Climate Change and Agricultural Production: Direct and Indirect Impacts (FAO 1996)*. [Online]. Available at: www.fao.org/docrep/w5183e/w5183e0c.htm [accessed 22 September 2011].

Roland-Holst, D. and Kahrl, F. 2008. *California Climate Risk and Response*, executive summary 3 (November, 2008). [Online]. Available at: http://are.berkeley.edu/~dwrh/CERES_Web [accessed 22 September 2011]

Ruhl, J.B. 2008, Making nuisance law ecological. *Case Western Reserve Law Review*, 58(3), 753–785.

Ruhl, J.B. 2008. Climate change and the endangered species act: Building bridges to the no-analog future. *Boston University of Law Review*, 88(1), 1–62. Available at: www.bu.edu/law/central/jd/organizations/journals/bulr/volume88n1 [accessed 22 September 2011]

Sax, J.L. 2010a. Land use regulation: time to think about fairness. *Natural Research Journal*, 50(2), 455–470

Sax, J.L. 2010b. The accretion/avulsion puzzle: Its past revealed, its future proposed. *Tulane Environmental Law Journal*, 23(2), 305–368.

Sigel, G. and Cassel, J.L. 2010. Climate change year in review: building foundations for change or just castles in the sand? *LexisNexis Emerging Issues*, 4838, 1–37. Available at: www.jenner.com [accessed 22 September 2011].

Strahilevitz, L.J. 2005. The right to destroy, *The Yale Law Journal*, 114, 781–854.

Taekema, S. 2004. *Understanding Dutch law.* Den Haag: Boom Juridische Uitgevers.

Tarlock, D. 2010. How well can water law adapt to the potential stresses of global climate change? *University of Denver Water Law Review*, 14(1), 1–46.

Titus, J.C. et al. 2009. State and local governments plan for development of the most land vulnerable to rising sea level along the US Atlantic coast. Environmental Research Letters 4. Available at www.iop.org/EJ [accessed 22 September 2011]

United States Climate Change Program 2003. *Final Report: Strategic Plan for Climate Change Program*, Ch. 6. [Online]. Available at: www.climatescience.gov [accessed 22 September 2011].

White, G. 1942. *Human Adjustment to Floods*. University of Chicago Dept. Geography Research Paper No. 29.

Wish, V. 2009. *Can Insurers Cope with Climate Change? (WWF-Allianz Study 2009)*. [Online]. Available at http://knowledge.allianz.com/climate/mitigation_adaptation/?98/wwf-insurance [accessed 22. September 2011].

Ziegler, E.H. 2006. Urban sprawl and sustainable development in the age of global terrorism. *William and Mary Environmental Law and Policy Review*, 30(1), 95–151.

Chapter 8

Government or Governance? The Challenge of Planning for Sustainability in the Ruhr

Michael Wegener

Sustainable development has a distinct spatial dimension. Decisions about the locations of workplaces, housing, retail, and leisure facilities have a significant impact on the achievement of sustainability goals. In recent times, under the influence of neo-liberal economic theories, more and more decision powers have been transferred to private actors and, in the interest of public participation, from higher to lower levels of government. The traditional model whereby government sets the framework for private decisions has been replaced by a flexible system of communication and adjustment between public and private actors called governance. From the point of view of sustainable spatial development, the trend from government to governance must be put into question. In a situation in which long-term ecological challenges, such as climate change and energy scarcity, are likely to exceed the current problem solving capacity of democratic decision structures, that capacity needs to be strengthened rather than further weakened. One way to achieve this is to make sure that decisions are made at the appropriate level of government at which not particular interests but the common welfare are pursued. This is not always the municipal level. The conclusion is that sustainable spatial development requires the reinforcement of democratic decision making at higher levels of government than the local level. This is demonstrated in the paper using the recent planning experience in the Ruhr agglomeration in Germany as an example.

This chapter takes up the distinction of this book between two ways of spatial planning: by legal regulation or by agreements between public and private actors (see also the Introduction by Hartmann and Needham); and it complements this distinction in two dimensions: power and spatial level.

In terms of power it distinguishes between the traditional model of *government* in which elected bodies set the framework for private decisions and *governance* as a flexible system of communication and agreements between public and private actors. Both models of planning work with laws and property rights, but they differ in the way these are generated and applied. It is argued that under the influence of neo-liberal economic theories, more and more decision powers have been transferred to private actors.

In terms of spatial level, it distinguishes between top-down and bottom-up approaches to planning and notes a trend of delegating more and more responsibilities to local levels in the interest of public participation. It is argued that this decentralization of decisions is in contrast to the extension of problem spaces through globalization, European integration and modern transport technologies.

It is the main hypothesis of this chapter that these two trends in planning, from public to private actors and from higher to lower spatial levels, are in conflict with new challenges for spatial planning.

New Challenges

There are numerous new challenges for spatial planning. Global warming, receding ozone layers, maritime pollution, destruction of tropical rain forests, decreasing variety of species and depletion of fossil fuels and other non-renewable resources are only a few keywords characterizing the ecological consequences of our growth-oriented way of life. For a number of years now, attempts are being made to find solutions to these problems using the principle of sustainable development.

Of the new challenges, the reduction of greenhouse gas emissions to mitigate global warming is the most recent and most demanding. The Kyoto Protocol of the United Nations of 1997 defined targets for the reduction of greenhouse gases. According to this Protocol, world-wide greenhouse gas emissions are to be reduced by 5.2 per cent until 2012 compared with the level of 1990. For the affluent countries of Europe this implies an average reduction by eight per cent, whereas no reduction targets were given for developing countries.

Under the impression of growing certainty of the threats of climate change, the heads of state of the European Union in March 2007 signed a declaration that by 2020 their countries achieve 20 per cent less energy consumption, 20 per cent renewable energy, and 20 per cent fewer greenhouse gas emissions compared to 1990 (and 30 per cent if other developed countries cooperate). At the G8 Summit in L'Aquila in July 2009 the political leaders agreed that the rich countries must reduce their greenhouse gas emissions by 80 per cent until 2050. Despite these commitments, the United Nations Climate Conference in Copenhagen in December 2009 turned out to be a sad demonstration of the 'tragedy of the commons', the over-use of free common resources (Hardin 1968).

In Germany already in 1990 a commission of the Federal Parliament had demanded that the industrialized countries reduce their greenhouse gas emissions by 80 per cent until 2050 in order to allow the developing countries to advance their economies (Deutsche Bundestag 1990). In August 2008 Chancellor Merkel announced that Germany will reduce its greenhouse gas emissions by 40 per cent until 2020. In its strategy for sustainable development of 2002, the Federal Government set the target to reduce the allocation of land for buildings and transport from 130 hectares per day to 30 hectares (Deutsche Bundesregierung 2002).

If these ambitious targets are to be achieved, spatial planning will have to make its contribution.

Sustainable Development

Sustainable development has a distinct spatial dimension. The utilization of space by humans for production, reproduction, consumption and leisure determines the consumption of material and energy, the locations of land uses and the pressure on the natural environment. The spatial distribution of land uses determines the spatial division of labour and hence the degree of spatial interaction and in that way energy consumption and environmental impacts of transport.

Decisions about the locations of workplaces, housing, retail and leisure facilities and the spatial interactions between them have a significant impact on the achievement of sustainability goals. These decisions are largely made by private actors (firms and households), but to a substantial part also by public actors, from local governments to the European Union.

At the same time, all spatial processes have undergone a historically unique enlargement in scale. Because of increasing affluence and technological advances in transport technology, cities have expanded more and more into their hinterlands, with the consequence of ever longer commuting and shopping trips. Labour market regions and catchment areas of central facilities have grown to a multiple of those in pre-industrial times. Disappearing trade barriers and low transport costs have multiplied the share of goods shipped from far-away countries. With growing interconnectedness of cities and regions, problem space and decision space increasingly separate.

Despite this, in recent times, under the influence of neo-liberal economic theories, more and more decision powers have been transferred to private actors and, in the interest of public participation, from higher to lower levels of government. The traditional model of government setting the framework for private decisions has been replaced by a flexible system of communication and adjustment between public and private actors called governance.

From the point of view of sustainable spatial development, the trend from government to governance must be put into question.

History of Spatial Planning in Germany

To understand this, a short review of the history of spatial planning in Germany may be useful (Wegener 1999; 2008).

The development of spatial planning in Germany reflects the history of the Federal Republic. After World War II, cities were rebuilt almost without planning. During the Cold War, the term planning was stigmatized as an expression of authoritarian government control. Only under the influence of the reform policy of the US presidents Kennedy and Johnson was this taboo abandoned. In the 1960s, under a social-liberal government coalition, societal planning became accepted for policy making also in Germany.

At the same time, the critique of the poor quality of cities rebuilt without planning increased (Mitscherlich 1965). It became obvious that to create high-quality residential and workplace environments, more comprehensive knowledge was needed than was taught at architectural and civil engineering schools. Following British and US examples, interdisciplinary planning schools were established at the universities of Berlin, Dortmund and Kaiserslautern. Under the influence of the then current paradigms of political economy and systems theory, comprehensive visions of the ideal spatial organization of society were developed. Particularly influential was the transfer of systems theory to social theory by the sociologist Luhmann. According to Luhmann, society is an open cybernetic system which survives against its environment by the selection of appropriate action. The reduction of complexity by stabilizing the difference between itself and its environment is the raison-d'être of a social system by which it distinguishes itself from biological systems (Luhmann 1966). Planning, and also spatial planning, as a means to reduce complexity, is therefore a kind of *system rationality* by which a social system maintains its existence (Wegener 1999).

But the 'planning euphoria' of the 1960s was shattered by the economic crises of the 1970s and 1980s. In the 1990s, the neo-liberal economic doctrine originating in the United States and Great Britain spread also to Germany and led to a gradual withdrawal of the state from spatial planning: from grand designs to small, manageable and correctable incremental improvements, and from hierarchical top-down planning to voluntary cooperation between public and private actors termed multi-level governance.

It was again Luhmann who called attention to the possibility that the democratic decision procedures practised today might not be able to respond adequately to the new ecological challenges (Luhmann 1986, 14):

> Thus the exposure to ecological self-endangerment remains within the context of the possibilities of evolution ... The possibility also exists that systems act on their environment in such a way that they cannot exist in this environment later on.

There are many historical examples of originally prosperous human societies that disappeared because they over-exploited their ecological resources (Diamond 2005). Does the slowness of ecological processes exceed the problem processing capacity of democratic societies? The procedures of modern democracies were developed in the eighteenth century when protection against dangers within one election period was sufficient. But is the response scheme of democratic societies also sufficient for planning for future generations? If not, what are the consequences for spatial planning?

Spatial Organization of Planning in Germany

The small-scale territorial organization of the Federal Republic of Germany is a fortunate heritage of the medieval division of Germany into many small states. Despite the subdivision into provinces, districts and counties following rational principles after the collapse of the Napoleonic Empire, and despite numerous functional and territorial reorganizations after World War II, important elements of the decentralized spatial structure dating back to the Middle Ages have remained intact, with the effect that Germany's urban hierarchy is today more balanced than those of most other European countries. This is one of the reasons for the relatively strong position of cities in the political system of the country.

Already in the 1970s there was a debate about the most appropriate spatial organization of spatial planning. Frido Wagener argued that the non-compatibility of planning space, administration space and decision space is an impediment for long-range development planning, and he proposed a five-level spatial planning hierarchy analogous to the federal system of Germany (Wagener 1970). More recent tendencies in administrative science and practice, however, have reinforced the responsibilities and decision powers of local governments in order to reduce bureaucracy and strengthen citizen participation (Landesregierung NRW 2003).

These tendencies affect spatial planning at all levels of the federal system. The reform of federalism of 2006 revoked the right of the Federal Government to define the principles of national spatial planning. Instead, both the Federal Government and the Federal States (*Länder*) are now entitled to make plans for the spatial organization of their territory, and the *Länder* have the right to deviate from national plans (ARL 2008). In several of the *Länder*, state planning was transferred to the intermediate level of government districts, as in North Rhine-Westphalia, or even to the local level, as in Lower Saxony (Bogumil et al. 2004).

The question is whether this decentralization of decision powers takes account of future challenges for the spatial organization of society.

Voluntary Cooperation

If Luhmann is right that future ecological challenges might exceed the problem processing capacity of today's democratic structures, that capacity needs to be reinforced rather than further weakened. One precondition for this is the greatest possible correspondence of problem and decision space. Due to the evolution of transport systems, however, the problem space of sustainable spatial development has long exceeded the jurisdiction of local communities.

The growing influence of strong private stakeholders on local government decisions and a tax system rewarding competition between municipalities rather than cooperation has led to urban sprawl and more unsustainable transport, in particular whenever under the motto 'less government, more market' regional planning above the local level has been disempowered or entirely abandoned.

Benjamin Davy, referring to Thomas Hobbes (1651), argues that the natural state of the world is 'war of everyone against everyone' (Davy 2004). To overcome this, there are three ways:

- to conclude a social contract in which power is voluntarily transferred to one or more individuals for rational and fair decisions about the distribution of goods,
- to leave decisions to the 'invisible hand' of the market, or
- to solve conflicts between actors by voluntary cooperation.

Davy illustrates the principle of voluntary cooperation by the 'prisoner's dilemma' between two municipalities deciding about new commercial or industrial areas: If municipality A plans a large area and municipality B a small one, A wins at the expense of B. If both municipalities plan large areas, both lose. If both cooperate and plan small areas, both win, but less.

Benz et al. (1992) discuss the problems of voluntary cooperation systems as follows: Pareto solutions in which all win are not problematic but rare. More frequent are Kaldor solutions by which inequality is accepted if the welfare gains are large enough to compensate the losers, for instance by payments or arrangements over time. The transaction costs of cooperation systems in terms of time and costs of negotiations are high. Without confirmation by an elected body, negotiated solutions lack democratic legitimation. In multilateral cooperation systems, the transactions costs multiply: the number of options to compare and the risk of cumulative vetoes grow with the number of cooperation partners, and compensations and arrangements over time become more difficult. If private cooperation partners are involved, the legitimation problem frequently becomes impossible to solve. Benz et al. conclude that cooperation is a necessary element of social life, but that multilateral cooperation systems are less efficient than multi-level majority systems. Also, cooperation systems are not more democratic than majority systems, and they imply serious legitimation problems, in particular when private actors are involved.

The conclusion for spatial planning is that, for the achievement of long-term goals affecting the whole society and not only individual municipalities, voluntary cooperation between competing municipalities is necessary but not sufficient, and democratic decisions at a spatial level above the local level are required – more government, less governance.

Example Ruhr Area

These relationships will be demonstrated using the spatial development in the Ruhr agglomeration in Germany as an example (Wegener 2009).

From the beginning of industrialization, spatial development in the Ruhr Area has been determined by the interests of large industrial corporations and the competition between the Ruhr cities. The result of 150 years of urbanization under these conditions is today's settlement structure with its disorganized, almost random pattern of vacant industrial sites, transport corridors and housing areas, and the continuing urban sprawl and reduction of open space. What is missing are a common spatial vision for a sustainable spatial development of the Ruhr area and the decision structures necessary for its implementation.

The Ruhr area was once a pioneer in regional long-term spatial planning. The 'General Settlement Plan' (*General-Siedlungsplan*) of 1912 for the western part of the Ruhr by the chief planning officer of Essen, Robert Schmidt, was an early example of comprehensive, long-term regional planning that won international reputation (Schmidt 1912). Its tradition was continued in the 1960s for the whole Ruhr area by the Association of the Ruhr Coal Mining Area Cities (*Siedlungsverband Ruhrkohlenbezirk*). The regional development plan of 1966 was the first legal regional plan in the Federal Republic; and at the same time the last one for the Ruhr (Petzinger 2008).

The *Siedlungsverband Ruhrkohlenbezirk* was founded already in 1920 and, since 1979 as *Kommunalverband Ruhrgebiet*, it is the oldest association of cities in Germany. Since 2004 the *Regionalverband Ruhr* is its successor. The regional parliament consists of representatives of eleven cities and four rural counties (*Kreise*) with 42 local governments. Membership is voluntary. In 2008, attempts by the county of Wesel and the city of Hagen to leave the *Regionalverband* failed only because they did not meet the required two-thirds majority. The cities of Dortmund and Hamm have repeatedly announced their intention to leave the association.

Until very recently there was no regional planning authority for the whole Ruhr area. Since the consolidation of local governments in 1975, regional planning in North Rhine-Westphalia has been in the responsibility of the five government districts (*Regierungsbezirke*) of the state. The Ruhr area is part of three of them: Düsseldorf, Münster and Arnsberg (see Figure 8.1). Each of them has set up its own regional development plan (*Gebietsentwicklungsplan*).

However, since the revision of the State Planning Law (*Landesplanungsgesetz*) of 2005, the regional plans are no longer binding for local governments. Three or more municipalities may issue a common regional land use plan which combines the functions of both a regional plan and a local land use plan.

For a long time, all efforts to establish a common regional plan for the Ruhr area have failed because of the resistance of local governments. Instead there has been a multitude of bilateral and multilateral cooperation initiatives between changing coalitions of local governments.

In a research project together with the Faculty of Spatial Planning of the University of Dortmund in 2003, the eight largest cities in the Ruhr area, Bochum, Dortmund, Duisburg, Essen, Gelsenkirchen, Herne, Mülheim and Oberhausen, formed the Ruhr Region of Cities 2030 (*Städteregion Ruhr 2030*) to deliberate together about the future of the region. In 2007 the cities of Bottrop, Hamm and Hagen joined. The objective of the cooperation was to overcome the narrow parochialism of the individual cities without losing their own interests out of sight.

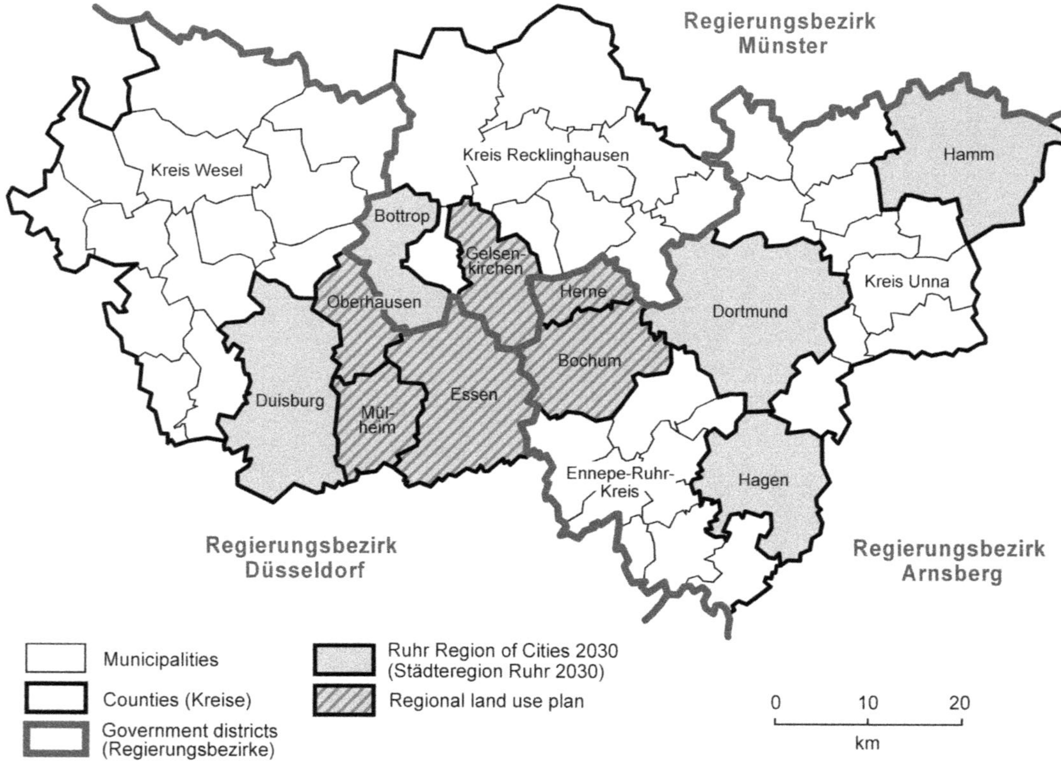

Figure 8.1 Fragmentation of regional planning in the Ruhr

Of the eleven cities, six – Bochum, Essen, Gelsenkirchen, Herne, Mülheim and Oberhausen – in 2005 established a planning association to develop a regional land use plan. The draft plan published in 2007 was criticized by a report commissioned by the state government (Greiving et al. 2008). The experts argued in that report that the plan assumes functions of regional planning in a part of the region, and so limits the freedom of future plans for the whole Ruhr area, but that at the same time its spatial resolution of five hectares is too coarse for local land use planning to have a binding effect for the municipalities.

In addition, there is a great variety of informal cooperations. One example is the draft of a 'Masterplan Ruhr' presented in 2006 by eight cities, Duisburg, Oberhausen, Mülheim, Essen, Gelsenkirchen, Herne, Bochum und Dortmund, with chapters on housing, urban design, and waterfront development. According to some participants, these informal cooperations increasingly lead to formal results, but at the risk of getting lost in too many negotiations (Tönnies 2008).

A good example of the problems of voluntary cooperation is the regional retail concept for the eastern Ruhr area. The concept was commissioned in 2000 by 21 municipalities (Econ-Consult 2000). Its general recommendations were to strengthen existing retail centres and to restrict retail development at non-integrated greenfield locations. In addition, recommendations were made for specific types of retail. Local food shops should be preserved as much as possible. Shops for clothing, shoes, toys, electronics, books, etc., should be allowed only in city or neighbourhood centres. Do-it-yourself and garden centres should only serve the local demand of a particular municipality. Only furniture shops were allowed to serve the demand of the whole region. According to the participating cities, the regional retail concept has helped to protect existing shopping areas and to prevent new non-integrated greenfield developments (Stitz 2007). In 2007 the concept was updated to take account of new developments (BBE Unternehmensberatung 2007). Despite this, in May 2008 a do-it-yourself superstore with 20,000 square metres of retail space was opened at a motorway exit between Dortmund and Witten without public transport access, grossly violating all principles of sustainable spatial planning.

In 2009, regional planning in North Rhine-Westphalia was again reorganized. To strengthen the international competitive position of the 'Ruhr Metropolis', the responsibility for regional planning in the Ruhr area was transferred to the Regionalverband Ruhr (Lageman et al. 2008). The Regionalverband had already in recent years identified objectives of regional planning to be taken into account in local land use plans. Since 2005, a 'Masterplan Spatial and Settlement Structure' for the Ruhr area has been under development (not identical with the 'Masterplan Ruhr' mentioned above). So far, however, the plan covers only the northern, mainly rural parts of the Ruhr area. It remains to be seen how conflicts with the regional land use plan of the six cities mentioned above, and with the other planning documents generated by the core cities of the Ruhr, will be resolved.

Conclusions

In Germany, Federal and *Länder* governments have increasingly withdrawn from national and regional planning and delegated responsibility for spatial development to local governments or private actors, with the aim to reduce bureaucracy and strengthen citizen participation.

From the point of view of sustainable spatial planning, this trend from government to governance must be put into question. In a situation in which long-term ecological challenges, such as climate change and energy scarcity, are likely to exceed the short-term problem-solving capacity of democratic decision structures, that capacity needs to be strengthened rather than further weakened – more government, less governance.

As long as all structures relevant for the relations between municipalities, in particular the influence of private stakeholders and the system of municipal taxes, reinforce competition rather than cooperation between municipalities, all efforts to achieve the objectives of sustainable spatial development, such as conservation of open space and reduction of car traffic, by voluntary cooperation between municipalities will fail. The most effective policies to achieve sustainable spatial development, policies that make mobility more expensive, can be implemented anyway only at the national or European level.

The conclusion is that sustainable spatial development requires the reinforcement of democratic decision making at the lowest possible level of government at which not particular interests but the common welfare are pursued. For an increasing part of spatial planning decisions this will be a higher level of government than the local level.

References

ARL – Akademie für Raumforschung und Landesplanung 2008. Zum Entwurf des Raumordnungsgesetzes 2008. Stellungnahme des Ad-hoc-Arbeitskreises 'Novellierung des ROG' der Akademie für Raumforschung und Landesplanung. *ARL Nachrichten*, (4), 3–6.

BBE Unternehmensberatung 2007. *Regionales Einzelhandelskonzept für das Östliche Ruhrgebiet und angrenzende Bereiche*. Fortschreibung 2007. Köln: BBE Unternehmensberatung.

Benz, A., Scharpf, F.W. and Zintl, R. 1992. *Horizontale Politikverflechtung. Zur Theorie von Verhandlungssystemen*. Frankfurt/New York: Campus Verlag.

Bogumil, J., Reichard, C. and Siebart, P. 2004. *Gutachten zur Verwaltungsstrukturreform in NRW*. Schriftenreihe der Stiftung Westfalen-Initiative 8. Ibbenbüren: Ibbenbürener Vereinsdruckerei GmbH.

Davy, B. 2004. *Die Neunte Stadt. Wilde Grenzen und Städteregion Ruhr 2030*. Wuppertal: Müller+Busmann.

Deutsche Bundesregierung 2002. *Perspektiven für Deutschland. Unsere Strategie für eine nachhaltige Entwicklung*. Berlin: Presse- und Informationsamt der Bundesregierung.

Deutscher Bundestag 1990. Dritter Bericht 'Schutz der Erde' der Enquête-Kommission 'Vorsorge zum Schutz der Erdatmosphäre'. *BT-Drucksache 11/8030*, 855. [Online]. Available at: http://dip.bundestag.de [accessed 9 June 2011].

Diamond, J. 2005. *Collapse. How Societies Choose to Fail or Succeed*. New York: Viking Press.

Econ-Consult 2000. *Regionales Einzelhandelskonzept für das Östliche Ruhrgebiet und angrenzende Bereiche*. Köln: Econ-Consult Wirtschafts- und Sozialwissenschaftliche Beratungsgesellschaft.

Greiving, S., David, C.-H. and Blotevogel, H.H. 2008. *Evaluierung des Planungsinstruments Regionaler Flächennutzungsplan*. Dortmund: plan+risk consult.

Hardin, G. 1968. The tragedy of the commons. *Science*, 162/3859, 1243–1248.

Hobbes, T. 1962 [1651]. *Leviathan or the Matter, Forme and Power of a Commonwealth Ecclesiastical and Civil*. London: Collins.

Lageman, B., Neumann, U. and Schmidt, C.M. 2008. Einheitliche Regionalplanung für die Metropole Ruhr: ein Erfolgsmodell? *Raumforschung und Raumordnung*, 66(1), 63–75.

Landesregierung NRW 2003. *Zukunft des öffentlichen Dienstes – öffentlicher Dienst der Zukunft*. Bericht der Kommission 'Zukunft des öffentlichen Dienstes – öffentlicher Dienst der Zukunft'. Düsseldorf: Landesregierung NRW.

Luhmann, N. 1966. Politische Planung, in N. Luhmann (ed.), *Politische Planung*. Opladen: Westdeutscher Verlag, 66–89.

Luhmann, N. 1986. *Ökologische Kommunikation. Kann die moderne Gesellschaft sich auf ökologische Gefährdungen einstellen?* Opladen: Westdeutscher Verlag. English: *Ecological Communication*. Cambridge: Polity Press, Chicago: University of Chicago Press, 1989.

Mitscherlich, A. 1965. *Die Unwirtlichkeit unserer Städte. Anstiftung zum Unfrieden*. Frankfurt: Suhrkamp Verlag.

Petzinger; T. 2008. Metropolregion Ruhr – Kern und Rand. Regionalplanung im Ruhrgebiet. *PlanerIn*, (1), 8–10.

Schmidt, R. 1912. *Denkschrift betreffend Grundsätze zur Aufstellung eines General-Siedlungsplanes für den Regierungsbezirk Düsseldorf (rechtsrheinisch)*. Essen: Stadt Essen.

Stitz, J. 2007. Kooperationsprojekt 'Regionales Einzelhandelskonzept'. Ein Erfahrungsbericht aus der kommunalen Praxis im östlichen Ruhrgebiet. *RaumPlanung*, (134), 194–198.

Tönnies, G. 2008. Regionale Kooperationen in Nordrhein-Westfalen. *ARL Nachrichten*, (4), 37–38.

Wagener, F. 1970. Von der Raumplanung zur Entwicklungsplanung. *Deutsches Verwaltungsblatt*, (3), 93–98.

Wegener, M. 1999. Raumplanung als Systemrationalität – oder die Rettung der Raumplanung durch die Ökologie, in K.M. Schmals (ed.), *Was ist Raumplanung?* Dortmunder Beiträge zur Raumplanung 89. Dortmund: Dortmunder Beiträge zur Raumplanung 89, University of Dortmund, 165–172.

Wegener, M. 2008. Zurück zu den Anfängen. *Informationen zur Raumentwicklung*, (11/12), 914–918.

Wegener, M. 2009. Nachhaltige Raumentwicklung als staatliche Aufgabe, in D. Schimanke (ed.), *Verwaltung und Raum. Zur Diskussion um Leistungsfähigkeit und Integrationsfunktion von Verwaltungseinheiten.* Baden-Baden: Nomos Verlag, 63–73.

PART IV
Coping with Uncertainty and Lock-ins

Chapter 9

Planning and Uncertainty: How Dutch Municipalities try to Reduce Uncertainty on the Market for Industrial Properties[1]

Huub Ploegmakers and Erwin van der Krabben

Recently, it has been suggested that public policy should play a stronger role in reducing and containing uncertainty in land and property markets by making this an explicit policy objective. (Adams et al. 2003; Adams, Dunse and White 2005). Adams et al. (2005) consequently argue that the issue of uncertainty should be included when analysing the impact of public policy on land and property markets. It is conjectured that different forms and combinations of public policy have different impacts on uncertainty reduction, although this has not yet been rigorously explored empirically (for an exception see Allmendinger 2006). Adams and Tiesdell (2010) thus observe that an answer is still awaited to the 'important' question posed by Healey, namely:

> Do plans stabilize land and property markets, creating greater certainty, thereby reducing transaction costs and by, limiting the potential for 'over-building' in a property market, producing greater efficiency in the relation between supply and demand in the land markets? (1992: 13)

The same could apply for the likely impact of other than regulatory tools on uncertainty in land and property markets.

A problem for this type of study, which is brought forward by Adams et al. (2003), is that the containment of uncertainty has never been considered as an explicit objective of public policy for land and property development, since the main focus of policy has been on the management of supply and demand. Adams et al. therefore conclude: 'It is thus perhaps unfair to judge a system by objectives that it has never formally articulated for itself' (2003: 40).

This is true perhaps in the case of passive forms of public planning, by means of a framework of regulations. When, however a public agency itself is actively involved in buying, servicing and selling land, one might assume that

1 We are grateful to Edwin Buitelaar for his helpful comments on earlier versions of this chapter and for allowing us to make use of the results of a recent study by the Netherlands Environmental Assessment Agency (2009).

the containment of uncertainty is a crucial objective, since it is the public agency itself that bears the development risks. The way public agencies themselves form expectations of an uncertain future under these circumstances might thus shed light on the more general question of the impact of public policy on uncertainty. An example of this type of case is provided by the market for Dutch industrial land, where most development is initiated by public authorities themselves. Therefore, this paper will investigate how municipalities try to reduce the uncertainty about the future demand for industrial land. We do this by analysing how they actually make forecasts of demand and how these estimates influence their development decisions by means of structured and semi-structured interviews with municipalities (making use of the study by the Netherlands Environmental Assessment Agency (2009) mentioned in footnote 1).

In the next section the interrelationships between planning and uncertainty are examined in general. In section 2 the way expectations are formed about uncertainties in the markets for land and property is discussed by reviewing the literature and previous research. This is followed by an account of the public development process for industrial land in the Netherlands, including the way in which municipalities form expectations of future demand according to the relevant literature. Section 4 presents the results from the survey of practice and Section 5 concludes.

Planning and Uncertainty

In the question quoted above, Healey (1992) restricted the potential impact on uncertainty to that caused by development plans. It is argued here that the discussion should not be restricted to the development control function of planning. Development plans, legal agreements and financial incentives all come under the title 'planning controls', but all have different effects (Allmendinger 2006: 6). In this respect, Tiesdell and Allmendinger advocate 'deconstruct[ing] the notion of planning as a homogenous and regulatory activity' (2005: 58). They argue that the heterogeneous nature of planning should be emphasized, for instance by showing that land-use regulation is only one part of planning activity. For this purpose, they have developed a typology of planning tools (for slightly different typologies see Adams, Disberry, Hutchison and Munjoma 2002; Geuting 2007; Segeren, Verwest, Needham and Buitelaar 2007). This identifies four types of planning tools, characterized by how they intend to affect the decision environment of market actors:

- *Market shaping tools* such as development plans, which provide the overarching context within which market agents make decisions.
- *Market regulation tools* which seek to regulate and control market actions by defining the parameters (for example choices available) of the decision context.

- *Market stimulation tools* which change the contours of the decisions environment by making some actions more or less advantageous.
- *Capacity building tools*: although difficult to define precisely, these tools aim to 'build' the capacity of actors to identify and develop more desirable strategies.

Tiesdell and Allmendinger (2005) regard direct state interventions, including the provision of serviced building land and infrastructure, as a special type of market stimulation, apart from fiscal measures such as subsidies and tax breaks. By making serviced land available – creating a framework of property rights – public agencies try to encourage others, such as building developers, to take actions which meet the aims of the planning agency.

Few studies have empirically investigated the impact of these different planning tools on land and property markets. Jackson and Watkins (2007), although they did not use precisely the same typology, find that a proactive policy environment, with a strong emphasis on town-centre management and improvements to the shopping environment, will exert a significant influence on retail property values. They argue that there is little evidence of the perceived negative effects of the regulatory aspects of planning activity (also see Jackson and Watkins 2005). Jackson and Watkins focus on the impact of different planning tools on retail property values, but Allmendinger (2006) investigates the impact of different tools on a broader set of indicators, a set which captures a market actor's decision environment. He did this by means of a survey and focus groups amongst key actors in the development industry. These indicators include the impact of planning tools on supply and demand, on information, and also on the extent to which uncertainty is reduced. The findings indicate that market shaping and several stimulating tools (for example grants and incentives) are able to reduce uncertainty while regulation and taxation (stimulation) increase uncertainty. Finally, capacity building has no significant impact.

Allmendinger (2006) does not explicitly address which sources of uncertainty are affected by these various planning tools. The relation between public policy and uncertainty in land and property markets is typically discussed in terms of the consequences of development control. Developers' desires to build can be severely constrained due to time-consuming and uncertain processes involved when a planning authority decides whether a proposed development meets prescribed rules. Certainly in the UK context, no developer buying land speculatively without planning permission knows whether or not planning permission will be granted (Mayer and Somerville 2000; Ball, Allmendinger and Hughes 2009; Ball 2011). Mayo and Sheppard (2001) call this type of planning 'stochastic development control', as there is some randomness in the planning authority permitting a project. Leishman et al. (2000) refer to this as planning uncertainty. In addition, they distinguish a second type of uncertainty faced by developers. Key development variables such as construction costs, property values and interest rates may turn out differently from those which were expected at the outset of development.

One could add to these variables the 'general' uncertainty whether there is (still) demand for land and property at the time the development project is completed. This type of uncertainty concerning the balance between supply and demand is labelled as market uncertainty (Leishman et al. 2000). When public agencies want to limit the potential for over-building in order to prevent commercial developers collectively 'shooting themselves in the foot' (Healey 1992) it is this latter source of uncertainty that can be reduced. The way to do this is not so much by stimulating the demand, but by reducing or phasing the supply of land (in the case of market regulation tools, by limiting development plans). This paper focuses on the latter type of uncertainty in land and property markets, and on how public agencies themselves try to reduce it when they are actively involved in land development.

Expectations in Land and Property Development

Since it takes several years to complete a development project, agents involved in land and property development have a strong interest in forming accurate expectations of future conditions: their success depends on making good forecasts of the key development variables. The literature about land and property development addresses how commercial market agents form expectations. In this respect, two ways of forming expectations are distinguished (see for example Ball, Lizieri and MacGregor 1998; Wheaton 1999; Barras 2005; Fuerst and McAllister 2010). The first is derived from the behavioural assumption in macroeconomics that expectations are formed rationally (see Muth 1961; Lucas 1972). This hypothesis implies that expectations – informed predictions of the future – are formed by taking account of all available information, including an understanding of the underlying 'economic model'. Consequently, Ball et al. (1998) assert that agents in land and property development use their knowledge of the property market, of the wider economy, and of the best available theories about how those two function and interrelate.

When applied to property development, this assumption means, according to Wheaton (1999: 215), that agents perfectly understand the equations that describe market behaviour and thus can make correct forecasts of rents – conditional on a particular future demand variable. When this (exogenous) demand variable in the model is known with certainty, then the rational expectation is equivalent to perfect foresight. If an unanticipated shock occurs to the demand variable, its effect on other variables can be forecast, even though the shock itself came as a shock to the market agents. Market agents have perfect foresight with respect to the effects, but not the timing, of the shock (Wheaton 1999: 215). Ball et al. (1998) add to this that the longer a project lasts, the greater the error of the forecasts. But as the error is assumed to be random, the consequence is as likely to be an over- as an under-estimation of demand. Thus, agents do not make systematic mistakes in forecasting future demand. It has to be emphasized that the rational expectations assumption is an aggregate hypothesis that does not necessarily explain expectation forming by an individual market agent; such expectations may be subject to a greater error than the theory suggests. However, although agents' expectations may be individually wrong, they are correct on average.

The second set of behavioural assumptions might be termed irrational or myopic expectations. The assumption is that market agents make their decisions on the basis of the current values of the important variables which affect their decisions. According to Wheaton (1999) many kinds of such 'irrationality' can be conceived. For instance, current levels of the relevant variables can be extrapolated, but also the previous growth rates of those variables (expectations are formed by trend extrapolation). Ball et al. (1998: 153) distinguish between naïve and adaptive expectations. Economic agents hold naïve expectations when they expect tomorrow to be the same as the present and adaptive expectations when they gradually adjust their views to new experience based on past changes.

Wheaton (1999) argues that although irrationality of economic agents is generally dismissed, some form of irrationality may still be occurring in land and property development. More recent studies are inconclusive about this. For office development, Tsolacos and Mcgough (1999) have investigated whether predictions are characterized by rational expectations. Their model suggests that expectations relating to office building development are indeed rational. On the other hand, a study by Antwi and Henneberry (1995) which makes use of behaviouralist modelling, indicates that commercial developers' future expectations are conditioned by past experience, for example expectations are formed by trend extrapolation. In this respect, they report that developers make use of trend extrapolation especially during periods of extreme market change (also see Henneberry and Rowley 2002). Using a similar methodology, Leishman et al. (2000) find that amongst house builders, uncertainty results in conservative forecasts of price levels, and that the predictions of house builders are based on current values. Nevertheless, the estimates that house builders make are not far removed from expectations if they had had perfect foresight. Finally, research across 20 European office markets by Fuerst and McAllister (2010) reveals the existence of divergent expectations. In some markets, developers seem to form myopic expectations, whereas in other markets developers did not respond to very strong demand signals. That latter might indicate rational expectations because the demand increases were only temporary and soon to be reversed.

Public Intervention in the Dutch Industrial Land Market

Public Land Development

In the Netherlands, other than in most other countries, development takes place in two related but distinct stages, each within its own market: land development and property development. Land is often exchanged twice before construction takes place. First the initial owner (supplier) sells (agricultural) land to a land developer (demander). In the 'secondary' market of serviced building plots (Halleux 2008: 260; Van der Krabben and Buitelaar 2011), the land developers become the suppliers, while the demanders of land are commercial property developers (building for sale or lease), final users (building for own use) and in the case of the land market for housing: housing associations. Building plots for industrial uses are normally located on 'formal' industrial estates, and our study is restricted to such formal estates. These are sites which the local land-use plan (bestemmingsplan) designates for activities in the sectors commerce, manufacturing and commercial services. Sites that are designated exclusively for offices are usually not covered. Since land for industrial uses may also be acquired on locations outside industrial estates, this study does not cover the total supply in the industrial land market. However, between 60 and 70 per cent of all employment in the sectors manufacturing and transport is located on (formal) industrial estates (Weterings, Knoben and Amsterdam 2008). Furthermore, we assume that industrial land transactions – the acquisition of building plots on which construction still has to take place – occur only on designated industrial estates, because industrial real estate transactions in other urban areas concern mainly second-hand industrial properties. That is, undeveloped land is hardly developed outside formal industrial estates.

Because Dutch municipalities can act as a legal person they can themselves develop land. Since World War II, municipalities have pursued an 'active land policy' with respect to industrial land. And they continue to do so, in contrast to housing development, where an active municipal land policy has come under pressure in the last decade or so (Buitelaar 2010). An active land strategy implies that municipalities acquire (agricultural) land, service it, provide the necessary infrastructure and re-parcel it into building plots, before it is sold off to interested parties (Needham 1997). Municipalities bear the corresponding risks and reap any profits. This strategy has been institutionalized through municipal land development departments. On industrial estates, approximately 73 per cent of all available building plots are owned and supplied as serviced land by municipalities, while private developers and semi-public regional development companies own respectively 13 per cent and 14 per cent (Segeren, Needham and Groen 2005; Van der Krabben and Buitelaar 2011). In this respect, Dutch municipalities are demanders of raw building land and suppliers of serviced building plots. Building construction can start immediately on those serviced building plots.

The development process of an industrial estate often starts with the municipality acquiring agricultural land. A land-use plan is not a necessary prerequisite for land acquisition. Municipalities are used to buying agricultural land for future industrial land development, sometimes in very early stages of planning (if an opportunity presents itself) and even before the land is allocated for industrial use. There are no (legal) restrictions restraining them from doing this. Often they make use of this right to buy (agricultural) land, in order to keep it in stock until it is needed for new industrial estates. Alternatively, municipalities may decide to make a land-use plan first, before they start to acquire the land. The advantage of this strategy is that they can make use of their compulsory purchase powers. However, they also risk paying a higher price, because the owners may withhold land due to expected price increases. Eventually, contractors will be called in by the municipality to prepare the site for development. At the same time, the local authority arranges the servicing (roads, water and sewers, power, etc.). As soon as the land is ready for development, it can be sold or leased.

Building plots for industrial use are usually directly sold to the final user (firms in services and manufacturing sectors), who commissions a construction firm to build the new industrial property. Virtually no exchanges take place between property developers and final users (or investors). The supply of new industrial property 'for the market' is, therefore, very small (Needham and Kruijt 1992; Louw 2000). It is important to note that this division of development stages results in a division of market uncertainties, or maybe better of development risks, although the market uncertainties remain interrelated. The land developer bears the risk that the supply price of serviced building land changes during the development and that the demand for that land has changed by the time the scheme is completed (see Table 9.1 for the length of the development process of an industrial estate). The property developer, when making the decision to buy the serviced land, faces the risks related to the construction costs and property prices and the risk that there will be no or less demand for the property when he has completed it. The property developer might also decide to delay the development, when he faces the risk that no more building land will be available for sale when he wants it (but that risk is very small for industrial land in the Netherlands). If it is the final user who acquires the building plot, he bears the risks only of the construction costs.

Table 9.1 Time it takes to develop an industrial estate

Activity	Average time span
Planning procedure	2 years
Land acquisition	3–4 years
Site preparation	1–2 years
Total	6–8 years

Source: Louw 2000

Forecasting Demand for Building Land

The municipality faces market uncertainty; therefore it wants to forecast demand as accurately as possible. Also, it wants to reduce the uncertainties for actors in the market for industrial properties. In the literature, several methods are distinguished that might be used by municipalities in order to make predictions of future demand for industrial building land (see for example Knoben and Traa

2008; Louw, Needham, Olden and Pen 2009). Some of the methods used by these public agents are similar to what the literature on expectations by commercial market agents describes as 'rational expectations' and other methods resemble 'myopic expectations' as they are based on some form of trend extrapolation (see Section 2). The first, a procedure known as the land coefficient method, starts by making an economic model of the demand for serviced building land, and uses this model to make forecasts of future demand. A fixed relationship between employment and industrial land use (the land coefficient) is assumed. Therefore, as employment rises, the demand for land increases proportionally. This approach was first put forward by Bak (1961). The Business Location Monitor (BLM) ('bedrijfslocatiemonitor', see Netherlands Bureau for Economic Policy Analysis 1997) is the best known adaptation of the land coefficient method. The BLM was introduced by the national government at the end of the nineties as a refinement of the existing land coefficient models, since it was known that municipalities both under- and overestimated the real demand for industrial land – sometimes with severe consequences – due to the inaccuracy of the land coefficient models extant at that time.

The BLM works in the following way (see Schuur 1999; Needham and Louw 2006). First, employment in the whole of the Netherlands is forecast. For this purpose, sectoral employment scenarios for the Dutch economy are used. Second, these estimates are partitioned between regions on the basis of regional and sectoral employment data. For example, regions with a larger than average share in the manufacturing sector would benefit more from the estimated national growth rate of this sector than other regions. Third, the amount forecast for a region is divided into employment on industrial estates, port areas and elsewhere. Fourth, the expected employment on industrial estates is translated into land area, by means of land coefficients, a fixed number of square metres of land for each future employee. Land coefficients differ by region and by type of economic activity. Finally, the stock of land on existing industrial estates is subtracted from that estimate. This gives the forecasts of net new demand for industrial building land.

The land coefficient method is not without criticism. Some argue that employment growth is a poor predictor of demand for land; the variation of land coefficients within sectors is shown to be substantial; and the coefficients have been obtained from a survey that was held in 1998 among 1200 businesses which had recently moved and from even older data (Knoben and Traa 2008). Regarding the first, Knoben and Traa (2008) have, however, shown that regional employment growth strongly correlates with the amount of industrial land sold or leased per year (revealed demand). It should be noted, however that the direction of causality might also be the other way around. Furthermore, Needham and Louw (2006) argue that the model assumes that the situation on the land market will not change structurally. As a consequence, policies for industrial land from decades ago are 'locked' into the current models, and no adjustments are made for emerging public objectives, such as intensifying industrial land use[2] (Needham and Louw 2006: 84).

The BLM provides provincial planning authorities with forecasts of the future demand for industrial land. Provinces are supposed to use the forecasts as input for their long-term planning. For this purpose, the estimated figures are compared with the amount of available building land on existing industrial estates and the amount of land designated for industrial use in local land-use plans which have not yet been realized. In the case that demand exceeds supply, plans for additional industrial estates have to be made. Provinces have to specify in their structure plans where new industrial estates may be developed. Municipalities are supposed to translate these expansions into their land-use plans (Louw 2000). In practice, however, it often occurs that provinces, as well as lower governmental agencies, use their own methods to predict demand (Olden 2010). Needham and Louw (2006) assert that the sum of all the separate predictions is usually higher than the prediction made for the country as a whole. Likewise, Olden (2010) finds that on average the provincial predictions are 27 per cent higher than the national forecast. Because provinces allocate a generous amount of land for industrial use, the municipalities face hardly any restrictions on supplying land for industrial development (Van Dinteren, Posthuma and Bruin 2007; Olden 2007; Netherlands Court of Audit 2008; Louw et al. 2009; Olden 2010).

The forecasts derived from the BLM are of restricted value to individual municipalities, for two main reasons. Firstly, as mentioned above, only for the region as a whole can demand figures be obtained from these forecasts. The validity of the model diminishes considerably at the local level, because locationally specific conditions are omitted (Arts, Ebregt, Eijgenraam and Stoffers 2005; Olden 2010). However, various consultancy bureaus have developed models which do enable forecasting of demand at the local level. Some of these models are quite similar to the land coefficient model. Secondly, the land coefficient models used to estimate demand for industrial land, either at the local or regional level, focus only on long-term demand for industrial land: for example, the most recent publication of the BLM presents forecasts of demand for industrial land for the period up to 2040 (Arts et al. 2005). These projections thus give virtually no clues about the timing of new development. Knoben and Traa (2008) therefore propose using

2 Since 2008 municipalities are obliged by the national government to make a better use of the existing stock of industrial land and property in order to reduce the development of new industrial estates. This policy is based on a principle put forward by the Social and Economic Council of the Netherlands (SER): what is called the 'SER-ladder. This principle entails that new demand for industrial space should where possible be accommodated by redeveloping older industrial estates or by intensifying land use on existing areas. Only if both options are inappropriate, can the demand for space be accommodated by developing new sites. As a result of this policy, municipalities might be expected to pay more attention to the existing stock of industrial land and adjust their forecasts accordingly.

econometric models, which explicitly take account of changing market conditions: this would make the models more suitable for generating middle-range forecasts.

Because of those limitations of the BLM, municipalities use other methods in order to estimate (short-term) demand for industrial land. One of these – known as the firm consultation method – entails interviews or surveys of the expected growth of existing firms. In the Netherlands, surveys among firms are used to obtain information about their location and property preferences. These surveys are used by governmental agencies to investigate the demand for various types of commercial property. The information generally sought relates to the number of firms seeking to move into new premises (Louw 2011). Louw (2011) distinguishes between this type of forecasts and econometric models, because surveys are often used to give forecasts for the short term (3–5 years), while models are used for long term forecasts (10 year). The firm consultation method has several disadvantages (see for example Knoben and Traa 2008; Louw et al. 2009). Firstly, the focus is solely on firms that are already located in the municipality.[3] Secondly, the method measures 'stated preferences'. It is doubtful whether the firm's expectations regarding new investments are always realistic. For that reason, municipalities run the risk of overestimating demand for new industrial land. The third method is trend extrapolation of recent rate of sales of industrial building plots (see for example Knoben and Traa 2008; Louw et al. 2009). For this method, the average sales rate in the (recent) past is determined. This trend is then extrapolated over the requisite duration. Because this method is relatively straightforward, it is probably the prevalent method among local and regional authorities. Olden (2010) has found that four of the twelve Dutch provinces make use of trend extrapolation; even to formulate long-term projections of demand. The author also asserts that most municipalities use this method. There is another reason why municipalities could be expected to base their supply on the amount of industrial land sold. Many municipalities maintain an 'iron reserve': a buffer of building plots, already prepared and serviced, so as to accommodate unexpected changes in sales rates. This is the result of the municipality's aspiration to have building plots available 'on tap' (Needham 1992: 684). As a rule of thumb, the available stock of serviced land should be equivalent to two to five times the average amount of building plots sold per year (Netherlands Bureau for Economic Policy Analysis 1999; Van Dinteren et al. 2007; Olden 2010). Assessments of sales rates in relation to the available stock may thus exert a crucial influence on the decision to start a new development scheme. Two other methods are identified in the literature additional to those described above (see for example Knoben and Traa 2008; Louw et al.

3 Firms in the municipality or the region are consulted in order to make forecasts of demand in the municipality or region. This assumes that demand comes from firms already in the municipality or region. This assumption is realistic since previous research has shown that most companies relocate within the region (94 per cent) or within the municipality (73–75 per cent) in which they were already situated (Van Oort et al. 2007; Weterings, et al. 2008).

2009). Both methods estimate demand as a function of expected output levels and only the parameters of the models differ. These methods have not been put in practice yet so they are not discussed further.

How Do Municipalities Forecast the Demand for Industrial Land in Practice?

This section relies on findings from semi-structured interviews and structured questionnaires that were carried out with municipalities by the Netherlands Environmental Assessment Agency (2009): 24 municipalities for the interviews, 64 for the questionnaires.[4] The first question was how municipalities got *long-term forecasts* of demand for building land. More than half of the interviewed municipalities said that they did not make their own forecasts, but relied on existing predictions made at a higher scale. Those predictions could come not only from the provinces, for adjacent municipalities might decide to co-operate in order to co-ordinate planning of industrial estates.[5]

Of the interviewed municipalities that made their own estimates of long-term demand, only three used models similar to the land coefficient method. Most of the other municipalities which made their own predictions extrapolated trends in the sales rate in the (recent) past. In this respect, four of the municipal respondents pointed out that when they assess previous sales rates, they explicitly take into account the ambition to have a reserve of building plots. Those municipalities which base their estimates on regional or provincial forecasts said that they compare these forecasts with previous disposal rates within their municipalities. One interviewee said that the expectations based on recent sales rates did not correspond to forecasts at the regional level, so the municipality reserved more

4 All the respondents to the questionnaire and to the interviews were working on policy for industrial estates in the department of economic affairs or of land development (depending on the internal organization) within their respective municipalities. The research concentrated on municipalities in which a minimum of 100 hectares of industrial land had been provided on dedicated (industrial) estates up to 2007, as listed in the national database of industrial estates (IBIS). This was the case for 150 municipalities.

5 For this purpose a law has been implemented, which facilitates voluntary co-operation between municipalities, the 'Joint provisions act' ('Wet gemeenschappelijke regelingen' – Wgr). This law was, however, not very successful and for that reason another solution was tried. Seven regions were appointed were planning at the regional scale was desirable. At the beginning of 2006 these regions were replaced by so-called 'plus regions'. As from that moment these regions are obliged to make a regional structure plan which sets out the desired future development of the region (Needham 2007). Not all Dutch municipalities are included in one of these 'plus-regions'. So when municipal respondents indicate that they rely on regional predictions, they may also refer to estimates produced by regional bodies in which municipalities voluntarily co-operate under the regime of the Wgr, instead of forecasts provided by the 'plus-regions'.

land for industrial uses than allocated in the regional plan. Finally, one interviewee pointed out that they compared the regional predictions with the results from a survey amongst local firms. The questionnaire results give a similar picture. Only one out of four municipalities forecast long-term demand at the local level: when they do, they are usually the larger municipalities. Fourteen questionnaire respondents say that they rely on estimates derived from trend extrapolation, on information about the number of firms seeking to move, or on a combination of both. The remaining respondents (half of all the municipalities) say that they rely on predictions of demand made at the regional or even provincial level. Most of those latter municipalities also take into account the previous disposal rates and/or results from surveys amongst firms.

The second question was to what extent decisions to initiate new development schemes were informed by formal forecasts of *short-run demand* for building land. None of the municipalities used the land coefficient method at the time it was decided to make more industrial land available. This should not come as a surprise since the land coefficient method is useful for long term forecasts of demand. Furthermore, only six municipalities used formal forecasting strategies to form expectations about demand in the short term. The majority of municipalities base their decisions to make more land available on extrapolations of previous sales rates. Only one respondent referred to the use of surveys of local firms. Municipalities seem to favour more structural consultations of firms. In this respect, local economic affairs liaisons officers (called *bedrijfscontact-functionarissen*) provide information about location preferences of firms which are already located in the municipality or planning to locate there. Sometimes this information is accompanied by waiting listsNonetheless, the number of municipal respondents that rely on informal contacts is relatively small compared to those which extrapolate recent sales rates. Finally, two interviewees said that decisions to develop new land were informed by the long-term forecasts only. The questionnaire results reveal that only one out of three municipalities make use of formal forecasting techniques when they decide to make more land available. More than half of the municipalities commence new development on the basis of their long-term projections of demand.

The fact that both the interview and the questionnaire results show that the majority of municipalities do not make use of formal forecasting strategies when they decide to make more land available, can be related to their aim to always have a reserve of serviced building land. That is, the new additions to the stock of building land have to keep up with the amount of land that is sold or leased to interested parties: one hectare sold, one new hectare supplied. This would imply that decisions to make more land available are not so much informed by extrapolations of the most recent sales rates, but by current sales levels of building land (for example naïve expectations as described in Section 2). The interview evidence shows that decisions to make land available are based on the principle that the amount of land designated for industrial uses has to be sufficient to secure a steady supply of serviced and prepared building plots. The aim is to prevent shortages in the face of unexpected fluctuations in demand. One municipal interviewee explicitly stated that long-term demand projections are not made, because new development is adjusted to the current disposal levels. Thereby the basic principle was to maintain an 'iron reserve' in order to respond to unexpected fluctuations in demand. Another interviewee pointed out that the ambition of the municipality was to rebuild an 'iron reserve' which had been used up, due to constraints on industrial land supply within the municipal boundaries and long planning procedures. One of the respondents said that the amount of land held in stock should be about 40 to 50 hectares, taking into account an average annual sales rate of 7 to 9 hectares. Another said that the 'iron reserve' should be equivalent to four times the average sales rate.

There is a potential supply of land not just on newly developed industrial estates but also on *existing* ones. This potential supply consists not only of vacant properties which have been put on the market, due to closures, relocations or otherwise, and which could be put to immediate use or occupation. For there is in addition a substantial stock of vacant industrial land on industrial estates which has never been built upon. Within the questionnaire, 58 of the 64 municipalities claim that they monitor the supply of vacant land on older industrial estates. Larger municipalities do this more often than smaller municipalities. Nonetheless, most municipalities assume that only a small part of demand for industrial space can be absorbed by the existing stock. Therefore estimates of existing availability hardly affect their decisions to make more land available by developing new industrial estates. Furthermore, those interviewed said that they take into consideration only changes in the amount of vacant land on existing industrial estates. And the empty, built, floor space in existing properties is not taken into account at all.

Concluding Remarks

There has been an increasing interest in the role of planning in containing uncertainties. Can governments reduce the probability that market actors produce outcomes which they themselves do not want, such as too many or too few buildings, or the wrong type of buildings, or in the wrong locations? Can governments reduce the transaction costs which market actors make in their attempt to reduce such uncertainties? Could this be one of the advantages of public planning? The fact (noted by Adams et al. 2005) that developers' future expectations might be conditioned by past experience suggests that there might be a crucial role for public policy in reducing uncertainty in land and property markets. On the other hand, if public planning puts forward a framework for market actors, based on expectations of demand, and if those expectations turn out to be wrong, the result is worse than if that framework had not been used. That would be the same if public planning makes frequent policy switches as it struggles to keep up with market trends (as noted by Adams et al. 2005). This is a good reason for studying how public agencies forecast demand for land and buildings. The way that Dutch municipalities forecast the demand for industrial land is a good case to study this, in particular because those municipalities, in supplying serviced building land themselves, bear the risk of bad forecasts. The question posed in this chapter is: how do Dutch municipalities do that, and do they do it well?

On the whole, municipalities forecast *long-term* demand by extrapolating recent sales rates of building land forward. Decisions to make more land available are informed by both trend extrapolation and current sales levels. The latter can be explained by the aim to maintain an iron reserve: one hectare sold, one new hectare supplied. The importance of current sales levels in forming expectations is illustrated by the fact that between 1991 and 2007 the total amount of 16,750 hectares of building land that was made available was almost equivalent to the amount of land that was sold or leased to interested parties, namely 16,600 hectares (Olden 2010). The question is whether this is a good method for predicting demand? The use of trend extrapolation and current sales levels in forming expectations does increase the likelihood of making systematic forecasting mistakes, which particularly becomes apparent during turning points in the market (both forecasting methods fall under the heading of 'myopic expectations'). Since it takes several years to develop a site, schemes that were initiated during prosperous market conditions, but just before the downturn in the market, will be completed when actual demand is much lower than expected (these forecasts could be either based on current sales levels or extrapolation of recent sales rates). Exactly the opposite occurs at the end of a downturn in the market: the use of either trend extrapolation or current sales levels to form predictions induces municipalities to expect that future demand will be low and therefore no additional land has to be made available for sale. The result is that when sales levels eventually begin to rise, it takes some time before the supply of land is increased as well.

Thus, one would expect that during an upturn in the industrial land market the probability of overestimates and subsequent oversupply of land increases and vice versa during a down turn. Indeed in practice, municipalities have not succeeded in anticipating turning points in the past. During a downturn, the stock of industrial land grows initially. This became most marked during the 1980s recession. At the second half of the nineties the opposite occurred during the earlier stages of a period of economic prosperity. Municipalities made less land available for sale than they actually sold. As a consequence the total stock of land that was immediately available for sale dropped by more than 25 per cent (Olden 2010). Evidence is absent whether the stock of industrial land was also growing just after the current global economic and financial crisis, and the accompanying substantial drop in the demand for building plots. What is clear is that the economic crisis has already brought some of the larger Dutch cities into (deep) financial trouble because of substantial interest losses on their land development (Van der Krabben and Jacobs 2011). These financial losses occur, because municipalities have not been able to sell the expected amount of building plots in time.

The conclusion is that Dutch municipalities do not cope well with the uncertainties on the market for industrial land. Ironically, the disadvantages are not borne by the building developers and end users, as might be expected when public attempts to 'shape markets' fail. The reason is the particular division of labour in the Netherlands: the disadvantages (the risks) of these poor predictions are borne mainly by the municipalities themselves, and the developers and end user face fewer risks. If the division of labour was different (for example municipalities bear no financial risks), those poor municipal predictions would increase the uncertainty – and thus the risks – faced by market actors.

References

Adams, D. and Tiesdell, S. 2010. Planners as market actors: Rethinking state–market relations in land and property. *Planning Theory and Practice*, 11(2), 187–207.

Adams, D., Disberry, A., Hutchison, N. and Munjoma, T. 2002. Land policy and urban renaissance: The impact of ownership constraints in four British cities. *Planning Theory and Practice*, 3(2), 195–217.

Adams, D., Dunse, N. and White, M. 2005. Conceptualising state–market relations in land and property: The growth of institutionalism–extension or challenge to mainstream economics? *Planning, Public Policy and Property Markets*, (37–55). Oxford: Blackwell.

Adams, D., Allmendinger, P., Dunse, N., Houston, D., Tiesdell, S., Townend, J. et al. 2003. *Assessing the Impact of Planning, Housing, Transport and Regeneration Policies on Land Pricing*. Universities of Aberdeen and Glasgow.

Allmendinger, P. 2006. Building on the past and present: Lessons for effective planning. *RICS Research Paper Series*, 5(12), 1–42.

Antwi, A. and Henneberry, J. 1995. Developers, non-linearity and asymmetry in the development cycle. *Journal of Property Research*, 12(3), 217–239.

Arts, P.H.A.M., Ebregt, J., Eijgenraam, C.J.J. and Stoffers, M.J. 2005. *De Vraag naar Ruimte voor Economische Activiteit tot 2040: Bedrijfslocatiemonitor*. Den Haag: Centraal Planbureau.

Bak, L. 1961. *Het Industriele Terreingebruik in Nederland. Een Planologische Facetstudie.* Den Haag: Staatsdrukkerij- en Uitgeverijbedrijf.

Ball 1998. Institutions in British property research: A Review. *Urban Studies*, 35(9), 1501–1517.

Ball, M. 2011. Planning delay and the responsiveness of English housing supply. *Urban Studies*, 48(2), 349.

Ball, M., Lizieri, C. and MacGregor, B.D. 1998. *The Economics of Commercial Property Markets*. London and New York: Routledge.

Ball, M., Allmendinger, P. and Hughes, C. 2009. Housing supply and planning delay in the south of England. *Journal of European Real Estate Research*, 2(2), 151–169.

Barras, R. 2005. A building cycle model for an imperfect world. *Journal of Property Research*, 22(2), 63–96.

Buitelaar, E. 2010. Cracks in the myth: challenges to land policy in the Netherlands. *Journal of Economic and Social Geography*, 101(3), 349–356.

Fuerst, F. and McAllister, P. 2010. Supply elasticities and developers' expectations: A study of European office markets. *Journal of European Real Estate Research*, 3(1), 5–23.

Geuting, E. 2007. Proprietary governance and property development: Using changes in the property-rights regime as a market-based policy Tool. *Town Planning Review*, 78(1), 23–39.

Halleux, J.M. 2008. Urban sprawl, urban containment and land management. A reflection on the concept of urban land supply, in M. Sitar (ed.), *Urban Future: Challenges and Opportunities for City and Region Developments*. Maribor: University of Maribor Press, 259–268

Healey, P. 1992. Development plans and markets. *Planning Practice and Research*, 7(2), 13–20.

Henneberry, J. and Rowley, S. 2002. Developers' decisions and property market behaviour, in S. Guy and J. Henneberry (eds), *Development and Developers: Perspectives on Property*. Oxford: Blackwell Publishers.

Jackson, C. and Watkins, C. 2005. Planning policy and retail property markets: Measuring the Dimensions of Planning Intervention. *Urban Studies*, 42(8), 1453–1469.

Jackson, C. and Watkins, C. 2007. Supply-side policies and retail property market performance. *Environment and Planning A*, 39(5), 1134–1146.

Knoben, J. and Traa, M. 2008. *De Bedrijfslocatiemonitor: Kritiek, Alterntieven en Aanpassingen*. Den Haag: Ruimtelijk Planbureau.

Leishman, C., Jones, C. and Fraser, W. 2000. The influence of uncertainty on house builder behaviour and residential land values. *Journal of Property Research*, 17(2), 147–168.

Louw, E. 2000. The production of business sites in the Netherlands. *Tijdschrift voor Economische en Sociale Geografie*, 91(1), 85–91.

Louw, E. 2011. *Surveys and forecasting industrial property demand*. Paper presented at the 18th Annual Conference of the European Real Estate Society.

Louw, E., Needham, B., Olden, H. and Pen, C.J. 2009. *Planning van Bedrijventerreinen*. Den Haag: Sdu Uitgevers (revised edition).

Lucas, R.E. 1972. Expectations and the neutrality of money. *Journal of Economic Theory*, 4(2), 103–124.

Mayer, C.J. and Somerville, C.T. 2000. Land use regulation and new construction. *Regional Science and Urban Economics*, 30(6), 639–662.

Muth, J.F. 1961. Rational expectations and the theory of price movements. *Econometrica: Journal of the Econometric Society*, 29(3), 315–335.

Needham, B. 1992. A theory of land prices when land is supplied publicly: The case of the Netherlands. *Urban Studies*, 29(5), 669.

Needham, B. 1997. Land policy in the Netherlands. *Tijdschrift voor Economische en Sociale Geografie*, 88(3), 291–296.

Needham, B. 2007. *Dutch Land Use Planning: Planning and Managing Land Use in The Netherlands, the Principles and the Practice*. Den Haag: Sdu Uitgevers.
Needham, B. and Kruijt, B. 1992. The Netherlands, in B. Wood and R.H. Williams (eds), *Industrial Property Markets in Western Europe*. London: E and FN Spon.
Needham, B. and Louw, E. 2006. Institutional economics and policies for changing land markets: The case of industrial estates in the Netherlands. *Journal of Property Research*, 23(1), 75–90.
Netherlands Bureau for Economic Policy Analysis 1997. *Bedrijfslocatiemonitor Terreinverkenning*. Den Haag: Sdu Uitgevers.
Netherlands Bureau for Economic Policy Analysis 1999. *Bedrijflocatiemonitor Regionale Verkenningen 2010–2020: in Gesprek met de Regio's*. Den Haag: Netherlands Bureau for Economic Policy Analysis (Centraal Planbureau).
Netherlands Court of Audit 2008. *Herstructurering van Bedrijventerreinen*. Den Haag: Netherlands Court of Audit [Algemene Rekenkamer].
Netherlands Environmental Assessment Agency 2009. *De Toekomst van Bedrijventerreinen: van Uitbreiding naar Herstructurering*. Den Haag: Netherlands Environmental Assessment Agency [Planbureau voor de Leefomgeving].
Olden, H. 2007. *Het Onderbouwen van Nut en Noodzaak van Bedrijventerreinen; een Kritische Beschouwing over de Praktijk van Marktonderzoek*. Utrecht: STOGO onderzoek+advies.
Olden, H. 2010. *Supplied Out Of Stock: The Overrated Role of Serviced Land as a Location Factor in the Planning of Business Estates* (in Dutch). Utrecht: Proefschrift.
Schuur, J. 1999. *Industrial Land Use Planning in the Netherlands*. De Haag: Netherlands Bureau for Economic Economic Policy Analysis, CPB report 1999/4.
Segeren, A., Needham, B. and Groen, J. 2005. *De Markt Doorgrond; Een Institutionele Analyse van Grondmarkten in Nederland*. Rotterdam / Den Haag: Netherlands Institute for Spatial Research [Ruimtelijke Planbureau].
Segeren, A., Verwest, F., Needham, D.B. and Buitelaar, E. 2007. (Re-)designing markets for land use decisions: private initiatives in publicly determined context: lessons drawn from other policy fields. *TPR*, 78(1), 9–22.
Tiesdell, S. and Allmendinger, P. 2005. Planning tools and markets: towards an extended conceptualization, in D. Adams, C. Watkins and M. White (eds), *Planning, Public Policy and Property Markets*. Oxford etc.: Blackwell Publishing, 56–76.
Tsolacos, S. and McGough, T. 1999. Rational expectations, uncertainty and cyclical activity in the British office market. *Urban Studies*, 36(7), 1137–1149.
Van der Krabben, E. and Buitelaar, E. 2011. Industrial land and property markets: market processes, market institutions and market outcomes: The Dutch case. *European Planning Studies* (forthcoming).

Van der Krabben, E. and Jacobs, H. 2011. Public land development as a strategic tool for redevelopment: what can American cities learn from the Netherlands? *Town Planning Review* (forthcoming).

Van Dinteren, J.H.J., Posthuma, A. and Bruin, L. 2007. *Regionaal Bedrijventerreinenbeleid; een Inventarisatie*. Nijmegen: Royal Haskoning.

Van Oort, F., Ponds, R., van Vliet, J., van Amsterdam, H., Declerck, S., Knoben, J. et al. 2007. *Verhuizingen van Bedrijven en Groei van Werkgelegenheid.* Rotterdam: NAi Uitgevers.

Weterings, A., Knoben, J. and van Amsterdam, H. 2008. *Werkgelegenheidsgroei op Bedrijventerreinen.* Den Haag/Bilthoven: Planbureau voor de Leefomgeving.

Wheaton, W. 1999. Real estate 'cycles': some fundamentals. *Real Estate Economics*, 27(2), 209–230.

Chapter 10
The Dutch Office Space Tragedy: Unlocking the Lock-in

Leonie B. Janssen-Jansen

Planners' aims to develop sustainable cities often conflict with the priorities of land developers and consumers in the local government's decision-making arena. In balancing these demands, municipalities often fail to resist development interests, and they change the regulations, in particularly because they themselves compete with their neighbouring authorities for jobs and inhabitants. Although this is explicable from local government's individual rationality, it often results in suboptimal situations at the regional level. An example is the over-abundance of office space and planned new office capacity in the Amsterdam Metropolitan Area. This can be characterized as a lock-in situation in planning.

The objective of this chapter is to investigate to what extent innovative ways of using law and property rights could contribute to finding solutions for these collective shortcomings, with the Amsterdam Office Space Tragedy as an example. After elaborating on the way the Amsterdam local governments have failed to control land use, I will analyse the set of regulatory decisions that have been proposed to address the problem, for example through the use of transferable development rights. Although this is a variant on planning by property rights, is also has aspects of a more collaborative approach.

Attention will be paid to the regulatory roles of the different levels of Dutch governments in this respect. Further, the challenges which these proposals will pose for the property owners and for the local governments will be discussed. Finally, I will reflect on the question whether law and property rights are the best way to deal with these challenges, or whether other approaches might result in more effective solutions.

Office Development Bubble

During the economic boom in the 1990s and 2000s, an abundance of finance speeded up investments in Dutch urban development. Public and private initiatives in a strong economy pushed for large-scale and high-end office and housing developments. Particularly on the office real estate markets, 'develop as much as possible' seemed to be the credo for a long time. Interest rates were low, finance conditions were increasingly relaxed, demand was booming due to high economic growth, investors liked the safe investments in real estate much more than investing in the unstable stock market. As a result, the stock of offices was continuously being enlarged, often stimulated by the governmental bodies involved in development.

In the urban political arena, local governments have always balanced their formal regulatory tasks against their interests in the field of land development and employment growth (cf. Swyngedouw 2005; Downs 2005). Research has shown that, even in good economic times with high growth rates, city governments have not been able to effectively resist developers' interests, even when they are at odds with the wider public interest of sustainable communities (Peiser 1990; Leo 1997). In more difficult times, when competition for tax-paying households and businesses is increasing, city governments are pushed even further to make urban development decisions which are based on considerations which are political, competitive and financial rather than spatially responsible (see also Peterson 1981). In their pursuit of economic and job growth, city governments always seem to be market-friendly and to follow the developers, and to grant permits for new developments, often relegating statutory norms and procedures to a secondary and subordinated place. In addition, most Dutch local governments also have a financial stake in developing new offices, due to their direct-development strategies.

These direct-development strategies, called active land policies by the Dutch, have been used by many local authorities and they have had a major influence on the process of land development. The municipalities bought and sold land themselves as a means to achieve planning goals, in addition to making plans as the zoning authority. Through this system, Dutch planning could supply affordable land for public goods like social housing for a long time. As municipalities owned almost all land for development, any value increment remained in public hands and was reinvested in the overall quality of urban and regional areas. These development increments, together with the pursuit of job and population growth, have stimulated municipalities to keep using an active development strategy, instead of a more regulatory approach, even after the land market became more market-oriented in the early nineties, when market players also started to buy land for development and speculation. Many municipalities still own vast amounts of land, which they hope to sell to the market for development. Office development generates the highest return on investment in land. This has encouraged the growth of commercial real estate. The risk of active land policy is borne by the public bodies.

Recently, conditions underpinning urban and real estate development have changed. Not only has demand declined due to the economic crisis and changing demographics, and the financial resources have evaporated, but also the focus of the office users has recently shifted towards more sustainable and small-scale developments. As existing vacant office buildings can hardly accommodate these wishes, the tenants often decide to abandon these buildings after their existing rent period has expired, leaving them vacant. There is hardly any demand for reuse, and this increases the vacancy rate in a region. The tenants rent another office, often newly constructed because developers and investors are still eager to develop new commercial real estate, as this still is profitable for them; investors also still see the development of new, sustainable offices with a guaranteed client as an interesting investment object. Therefore, new office development continues, making older offices out-of-date even faster. By now, over 50 per cent of the vacant office stock in the Amsterdam region suffers from structural vacancy. Some locations, for example Diemen-Zuid, face vacancy rates up to 70 per cent by the end of 2011.

The office bubble is often accompanied by a retail bubble, industrial parks bubble and a housing bubble, as a result of past planning decisions based on growth-oriented perspectives. Facing a period of non-growth will challenge urban development enormously. As of now, the cities have hardly responded to this and just continue their existing policies (Janssen-Jansen 2010a; 2010b).

From that perspective, market initiatives for the construction of new offices are often still welcomed by cities, also because the latter increasingly welcome development that is more sustainable than the existing stock. From a more economic perspective, a company's decision to move to, or stay in, a region is also welcomed and stimulated, as it secures jobs within the municipal boundaries. In this sense, Dutch planning is quite market-oriented. Yet, in most areas it is only possible to develop new office space if this fits the zoning conditions: the area has to be zoned for office development. In Amsterdam and many other cities, an additional regulation exists: the land regulations require that new developments are only possible if the end user (for the first five years) is known. Without an end user, no building permit will be granted. This 'pre-rent' regulation applies to all new office development, independently of which actor is the land owner. The zoning plan often has to be adapted before a building permit can be granted. Changing the zoning plan is the responsibility of the municipality. If the zoning plan allows office development, municipalities can still require a known end user before granting a building permit. Since October 2010 an All-in-one Permit for Physical Aspects (omgevingsvergunning) is needed. The permit is covered by the Environmental General Provision Bill (Wet algemene bepalingen omgevingsrecht, Wabo).

With all municipal players in a region focusing on high quality office development in order to increase their local competitiveness and to profit from value increments as a result of their active land strategies, local competition is strong. However, as a result of all these 'develop as much as possible' strategies, the competitiveness of the metropolitan region as a whole decreases due to the oversupply of offices. Each actor is motivated to maximize its own 'use' of the available demand for office spaces, while the costs of over-exploitation are distributed between all those to whom this demand for office space is available. Here, these include not only the municipalities who decide to build additional offices, but also those who decide against that and also existing office owners (who face devaluation of their properties) and the inhabitants of the region who miss out on other potential land-use amenities and in the case of vacant offices, have to face the site problems.

Currently, about 14 per cent of all Dutch office buildings are vacant: over 7 million square meters (DTZ 2010). The overall office vacancy rate in Amsterdam hit 20 per cent long before the economic crisis (Janssen-Jansen 2010a, 2010b). This vacancy rate is among the highest in Europe (see Figure 10.1). Where 4–8 per cent can still be seen as frictional vacancy, a vacancy rate of 20 per cent indicates a structural problem. As mentioned above, over 50 per cent of the vacant stock suffers from structural vacancy.

Since 2002, the overall vacancy rate has not been below 9 per cent. Many office buildings often outdated and at relatively unattractive locations, are likely to become and remain obsolete. For the international competitiveness of cities, a structural oversupply is harmful (TNO 2009). It worsens the look and image of a city, which directly and indirectly affects the demand for new office space (Van Gool et al. 2007).

Because the municipal land development strategies mean that the municipalities carry the financial risks of the land development, some municipalities will face severe financial problems in the near future, and even more if they have been overoptimistic in planning and developing industrial parks, retail and housing too. As a consequence, they have an important reason for continuing their 'develop as much as possible' strategies because then they will sell their land stocks and thus bring in revenues. Most of the municipalities in the Amsterdam region, but also in other parts of the Netherlands, therefore have included new office capacity in their current plans, despite the high vacancy rates in their areas, thus prolonging the office bubble (Janssen-Jansen, 2010b). They hope to be able to sell land to

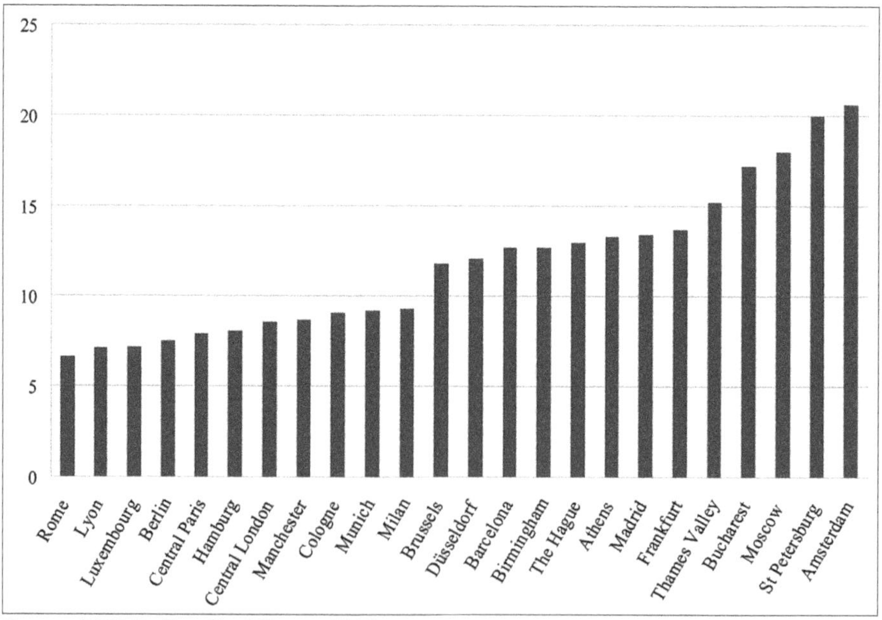

Source: BNP Paribas Real Estate, 2010

Figure 10.1 Office vacancy rates in European cities, Q2-2010

developers, lessening the burden of interest rates, and they hope that it might stimulate economic development and attract new companies. A classic collective action problem emerges (Ostrom 1990) and a clear lock-in situation. Such a collective action problem is about rational actors who, acting independently and instrumentally, cannot be counted on to commit resources to a common cause (Gordon 2006: 140).

Viewing it from a metropolitan perspective, a kind of 'tragedy of the commons'[1] evolves, as a conflict arises between the use of resources for individual interests and for the common good (Hardin 1968). Not only is the demand for office space by market actors often seen as a kind of resource for development by municipalities, but so is the open space that might be used for office zoning in a region also. And although the open space, at least in the Netherlands, is not freely accessible (as it is always one of the municipalities which has to decide whether or not to use the open space for another development), the total needed space for office development in a metropolitan region is finite. The demand for offices can in regional perspective be seen as a shared resource for development opportunities.

New buildings are being added to the supply, older buildings become redundant. To some extent, new uses in those old buildings will be possible, but in areas with high levels of office vacancies, like in the Netherlands, there is hardly any prospect for these older and less environmentally friendly buildings being used for new offices. From a planning perspective with ambitions for a sustainable urban future, a logical step would be to stimulate redevelopment of those sites. However, transformation and restructuring is a complex process involving many diverse stakeholders and high costs of land acquisition, land rehabilitation and construction. Even with high density plans for housing or offices, the units in redevelopment projects are relatively expensive. With an extremely weak demand in this higher end of the housing and business markets, decreased trust in the stable value of real estate, and the changed behaviour of financial institutions towards business investments and household mortgages, the return on investment in restructuring processes is, as a consequence, very uncertain for developers and investors. This discourages them from starting redevelopment at all. Yields on developing new offices are more certain. As said before, local governments, constantly competing for private investment, find it hard to resist new developments. Yet, at a certain moment, one has to acknowledge the problematic situation.

---

1 In the history of the theatre, a tragedy is a play with an inevitably fatal outcome for the protagonist. He 'is brought to ruin or suffers extreme sorrow, especially as a consequence of a tragic flaw, moral weakness, or inability to cope with unfavorable circumstances' (www.answers.com). A 'tragedy of the commons' is a concept in legal economics and game theory. It describes a social dilemma in which short-term self-interest conflicts with the group long-term interest (Janssen-Jansen, 2010). Here, a comparison with the classic tragedy of the commons will not suffice because the demand (the shared resources) is not endless, and some marginal costs exist. Both of these elements are not sufficiently taken into account.

The Amsterdam Metropolitan Area, suffering from high office vacancy rates, has been searching for solutions to the collective shortcoming of office development in its region. At first, in 2005, the municipalities were concerned with the enormous planned capacity (zoning) for new offices, but more recently attention has shifted to the problem of the vacant offices themselves, reinforced by policy attention at the national level and proposed regulatory decisions to address the problem. Due to demographic developments, and the introduction of more flexible forms of work, it is likely that the structural vacancy will grow, re-emphasizing the need for action.

The next section will elaborate on the way the Amsterdam region is dealing with the office overcapacity. Section three will deal with the proposed regulations and ordinances, for the existing vacant buildings, but also for preventing new situations of overdevelopment. The final section reflects on the question whether the use of law and property rights will result in effective solutions.

Regulating Office Capacity in the Amsterdam Metropolitan Area

Because of its societal impact, office supply is subject to urban planning and local spatial regulation. At the regional level, however there is in the Netherlands hardly any regulation, although relations between businesses, labour, banking and finance (major consumers of offices) are rescaling to regional and even international levels. In the Amsterdam region, this lack of regional office regulation has resulted in an oversupply of offices. In 2006, the vacancy was over two million square meters (23 per cent of the stock). The existing plans indicate that up to 2030, over eight million additional square meters are foreseen. According to estimates of supply and demand, by the end of 2005 sufficient regional office space had been planned for the next 83 years (Van der Plas 2006). According to calculations based on economic growth projections, the demand would be about 3.5 million m^2 until 2030. Although it is obvious that this total disregard of the user market for office space is undesirable, it took a number of years before the involved players felt obliged to search for a solution. None of the actors, neither market nor government, suffered from the problem. Until recently, most involved actors regarded the oversupply of office space as a market failure. They argued that the oversupply was a stage in the hog cycle[2] and they did not categorize it as a public

2 The phenomenon of cyclical fluctuations of supply and prices in livestock markets. For office development: in a phase of business upturn and development available space is absorbed quickly and new development is attractive for investors because of lower interests rates and higher rents, but the moment the new development comes on the market, real interests rise in response to the boom and the business circle turns downward. As vacancy rates rise, rents falter and new development is postponed, although for some time new development comes on the market as result of the pipeline effect. After this business downturn and overbuilding phase, the market should adjust with falling rents, which –

problem. In their opinion, empty offices were a problem only for their owners, the investors and developers who built them 'at risk' anticipating high user demand (Ploeger 2004: 140), not a problem for society as a whole.

However, the oversupply of offices in the Amsterdam Metropolitan Area has never been solely a result of market failure, but also of public failure due to the above described direct land development strategies, and the constant amendment of zoning regulations in order to accommodate 'market forces'. As we have seen above, municipalities compete with their neighbours to attract new businesses. New offices are often the most profitable elements in development, more profitable than for example housing projects, not only for the investors and developers, but also for municipalities, and even more so if the municipality itself is the landowner – which is quite often the case in the Amsterdam Metropolitan Area. Land disposals give Dutch municipal governments a strong position in urban developments in the Netherlands. Selling developable land is the main source of independent income for municipal governments (Janssen-Jansen and Salet 2009). Moreover, after realization of the real estate, municipalities earn property taxes yearly. Due to the Amsterdam land lease system (on land which is owned by the municipality), the municipality of Amsterdam also continues to collect property taxes on empty offices and thus does not really feel a need to change its planning policies. Changing the use from offices to, for example, residential forgoes income, because property taxes are lower when a building is in residential use. The property tax perspective adds another dimension to the embeddedness of this problem. The existing regulations reinforce the problem and the persistence of this lock-in in planning. None of the involved actors is prepared to react to the changing situation. All municipalities in the Amsterdam Metropolitan Area have counted on the income from new development in order to finance municipal services, including, for example, the restructuring of social housing areas (Janssen-Jansen 2006). The eight million square meters of planned office space were all linked financially to less profitable developments. In 2005, the mayors, aldermen and provincial deputies of the Amsterdam Metropolitan Area acknowledged that society as a whole suffers from the enormous oversupply of offices. In particular the areas with high vacancy rates had a poor quality, and could deter possible investors in the region. It could no longer be denied that the Amsterdam Metropolitan Area was in need of new approaches, and the local politicians began to search for regulatory decisions at the regional level. Firstly, they decided to create a platform to establish a new balance between the demand and the supply. This Platform *Bedrijven en Kantoren* (Businesses and Offices – Plabeka 2005) was established in 2005. The aim was to channel the inter-municipal competition and to focus on the quality of locations instead of quantity of offices constructed. In February 2007, the decision was made to eliminate 3.5 million m^2 of the 8 million meters planned capacity of offices; 1.5 million m^2 of office space

eventually – eases filling of vacancies when an upswing of the business cycle occurs (Ploeger 2004: 35–36).

planned for Amsterdam, 1 million m^2 for the municipality of Haarlemmermeer (which also includes Schiphol airport), and 0.5 million m^2 for the municipality of Almere The other municipalities were responsible for eliminating the remainder (0.5 million m^2) (Janssen-Jansen 2007). Following this decision, each municipality had to negotiate which zones designated for offices would be de-zoned; the zones on municipally owned land, or on privately owned land, sometimes already zoned for offices. This would imply removing future development possibilities. As this would be a down-zoning, for which compensation has to be paid, most of the municipalities chose to cancel the 'soft plans', and preferably the ones in which they themselves had only a small financial stake.

Although this cancellation of plans was seen as an important regional policy achievement by some (Haran 2010), from the moment of the agreement, it was immediately clear that the region had failed to find a real solution to the office vacancy problem (Janssen-Jansen 2007). All municipalities got a part of the – in their opinion – growing pie, but it was already clear that even with these cancellations of zoned office space there would still be an oversupply. Now, with only half of the pie left due to the economic crisis, the regional partners face difficult negotiations that will test the strength of the existing regional agreement, as it remains to be seen whether all players will stick to the agreements (Janssen-Jansen 2011). The Plabeka-II negotiations started in 2010, and although it can be argued that these negotiations would never have got under way without the first round, it is already clear that in a few more years, or even sooner a Plabeka-III round will be needed, and it is likely also that the planning for retail will then be included. The Plabeka-II round resulted in agreement on additional cancellation of plans, but the assumptions about office development in the Amsterdam region are still optimistic and based on a high growth scenario.

The platform continues to be informal in nature, with a pragmatic agenda, but without enforceable regulations that could prevent another such fiasco. The distribution of future office development is politically a difficult process, due to the various and conflicting, interests and preferences across the municipalities, which reinforces inter-municipal competition (c.f. Allmendinger and Haughton 2010).

The financial interests of the municipalities are enormous, as a result of which municipalities resist cancelling plans; they had already reckoned on the future income from land, had acquired land for possible future development and as a result have to pay interest charges for this land. Cancelling plans may cause severe financial problems for municipalities, in particular if also the private actors need to be compensated. As a consequence, all municipalities find it hard to decide to dezone office land, despite the agreements. In this way they prolong the problem.

Some Proposed Solutions

Fair and honest agreements about who develops where and how to change the agreements if markets change, could strengthen the potential of a region.

Reasoning from this perspective, a sustainable and healthy market for office space in a region requires even more collaboration than now exists in order to decrease the new supply of offices to suitable proportions. This collaboration would enable the problem to be explored more objectively, including, for example, the opportunity costs of (new) vacancies. At the regional level, several solutions have been put forward to regulate the supply. One of the ideas is to introduce a system of transferable development opportunities to balance the planned capacity for office developments between municipalities. These latter all perceive development of offices as a right, as we have seen above. A system of transferable development rights would consist of imposing quotas for office development; each municipality would acquire 'office rights' or office quota that it could use for its own development projects or sell to neighbouring municipalities. This transfer of 'office development rights' from one municipality to another might be preferable for the region as a whole and preferable to the involved municipalities. The local municipalities can decide on this among themselves by entering into contracts, which would include more or less top-down conditions about how the number of new office buildings should be related to the demand in a certain period. A regional authority or the province could take a role here. If a municipality wants to develop more than is agreed upon in the regional agreement and allocated through office development rights, it should pay a contribution to a regional fund that would be used for improvement of the overall spatial quality in the region. Another option is to 'buy' a part of the office space potential from another municipality. This can be a financial transaction, but can also concern an investment in the spatial quality. In both ways, either through buying rights from another municipality, or through a contribution to a regional fund, self-interest will be channelled to the shared interest. Every municipality will get, in a regional negotiation process, a share of development potential, with increased transparency about opportunities and developments. The municipalities involved might agree to have an outside supervisor to monitor the agreements (Janssen-Jansen and Salet 2009).

A major problem with such a system would be the decision about the amount of development to be allocated to each municipality. Should it be based on, for example, available land, economic interest, or the number of inhabitants? And of course, there should be a link to the demand in the region. Such a system would never resolve all competition issues, but might help to break the vicious circle. This trading system of office development rights might prevent (the increase of) future vacancies. Yet, it will not solve the 'dead office' problem, the existing vacant stock. Recently, other regulatory decisions have been proposed to address this.

Regulating Vacant Office Buildings and their Transformation

Many ideas, solutions and instruments have been introduced to address the problem of vacant and often obsolete office buildings. Some are repressive, like the introduction of a vacancy tax, aiming to give an incentive to office owners to restructure their building in an early stage. Other forms of regulation are oriented on stimulation, to encourage the redevelopment of vacant office buildings. The city of Amsterdam has recently shown interest in developing new instruments for the conversion of obsolete office buildings (Oudeman 2010), instruments which are additional to existing regulations, subsidies, orders, prohibitions and permissions. The facts that the city of Amsterdam follows an active land-use policy and that the local land lease system influences the development of land, enlarge the possibilities for changing the use of certain instruments in this municipality. Yet it has to be realized that, given the lock-in described above, the municipality also has a stake in not restructuring vacant office space, in particularly as this might take away initiatives for new office space development.

This section will focus on some of the municipal initiatives. Yet is must be noted that, at a national level, the fiscal regulations also have a major influence on prolonging the office vacancies, as owners get a tax reduction for vacant space, which could be an incentive to leave the building as it is. Furthermore, other fiscal regulations also prevent owners from cooperating in transformation initiatives. This is a huge problem for dealing with vacancies, but is beyond the scope of this chapter.

Although there has been quite a lot of attention at the national level in the Netherlands for the office vacancy problem, the right-wing government elected in 2010 decided to devolve both coordination and monitoring of spatial planning and development as much as possible to the provinces (VVD and CDA 2010). The provinces, the intermediate government level in the Netherlands, have now been given a supervising role in order to overcome the public failure of municipalities with regard to business parks and office development – overlooking the fact that the provinces have long had this authority, but have never used it to prevent vacancies by restricting the number of business parks, office sites etc. Further, planning does not stop at provincial boundaries. In addition, with the introduction of the 2008 Dutch Spatial Planning Act, the provinces themselves have become more actively involved in land development, sometimes even duplicating the active direct-development strategies of the municipalities. Thus, it remains to be seen whether the provinces are able to prevent the municipal pitfalls of business parks and office planning, or whether they will be able to deal effectively with the proposed regulation.

This remainder of this section will elaborate on some municipal initiatives, like the ideas for introducing a vacancy tax, more flexible zoning, mechanisms of urban land readjustment, or the office-for-office instrument. Some of them are part of the 'unorthodox' regulations recently initiated in the Netherlands to address the vacancy problem.

Vacancy Tax

Experiences in Amsterdam, where the land lease agreements have enabled imposing such a tax for decades, showed that introducing such a tax is impracticable, because of enforcement and monitoring, but also because of the inter-municipal competition. Municipalities have a stake in new office development, and they do not want to push away investors to other municipalities. Yet, with the introduction of the 'Anti-squatter Act' in 2010, all municipalities can impose vacancy taxes. From the wish to address the office problem, many municipalities are now drafting 'vacancy ordinances', including the City of Amsterdam (Municipality of Amsterdam 2011a). Although highly contested, one of the measures that are included in the ordinance is to fine owners of a building that is vacant for more than a year. It would be a fine of €7,500. Many involved actors only see this as a symbolic regulation, as €7,500 is nothing compared with the cost of a vacant office building, or the risks involved in restructuring the building. Again, doubts exist whether it is possible to enforce and monitor vacancies in order to fine the owner at all; what to do for example with partial vacancy?

Some politicians argue for reintroducing possibilities to squat a vacant building and abandoning the 'Anti-squatter Act', as they see this as way of activating owners to find ways to reuse their buildings. However, when squatting was legal in the Netherlands, for buildings that were vacant for more than one year, hardly any of the vacant offices got squatted, as they are on locations that are not of interests to squatters at all, in the outskirts of the cities. Only vacant offices in the inner cities were vulnerable to squatting. An initiative from concerned citizens in the municipality of Haarlemmermeer (Schiphol) to have a building squatted failed for the same reason.

As part of the vacancy ordinance that became effective on 1 July 2011, the municipality of Amsterdam also proposed that it should actively find new users for obsolete buildings, forcing the owners to cooperate. The municipality will actively monitor the results of the ordinance. It has been decided to evaluate this new instrument after 18 months.

Flexible Zoning and Rezoning

There also have been proposals for stimulating restructuring and transformation of obsolete office buildings, especially the ones that deteriorated and have hardly any prospects for reuse for office purposes. Demolition and new construction could be a possibility, but is often perceived as an economic loss by the owner. More likely is the transformation of a building, for example to creative working spaces, student housing, apartments, hotel etc. As these other functions are often not in conformity with the current zoning plan, many of the proposals for transforming a building need to be realized within the flexibility offered by the spatial planning law. Temporary use for 5 years is quite often allowed, but the business case for transformation for only 5 years is often hard to make: for example, the investments for changing an office building into student housing cannot be recouped in five years. Now, it is being debated whether to extend the term for temporary use to for 10 years, instead of five: this would help to make temporary transformations financially feasible. Yet, the question still remains: transformation to what? Given the demographic developments, it is questionable if the current shortage of student housing will last more than a couple of years. In addition, it is highly questionable whether students want to live in the areas where the vacancy rates are high.

Many initiatives for more permanent transformations also exist. Owners of an obsolete property, for example, can explore options to transform their building into a hotel. Amsterdam currently has a shortage of hotel rooms (Municipality of Amsterdam 2008). The municipality adopted policies to actively stimulate the transformation of vacant office space into hotels (Municipality of Amsterdam 2007; 2011b) and actively stimulates owners to participate in the rezoning initiatives. Sometimes this requires that the municipality be flexible with its own regulations about the availability of parking spaces. Owners of buildings and hotel operators start a transformation only if it is more profitable than leaving the building vacant. The municipality of Rotterdam also is actively stimulating owners to transform their buildings into housing, schools, hotels, shops or other functions (Municipality of Rotterdam 2011).

In 2010 the Netherlands enacted a special Crisis and Recovery act, containing literally hundreds of articles, all of which are meant to speed up decision-making on a wide variety of activities, to speed up projects and diminish delays in court. The *Crisis and Recovery Act*, as *ad hoc* legislation, includes several experimental instruments. The Act was drafted in a short time and is meant to be in effect for only five years, but it in June 2011 it has been proposed to adopt this Act permanently. This act should increase the flexibility in zoning and the possibilities for rezoning.

Urban Land Readjustment

Urban land readjustment is another initiative that is gaining attention as a way of addressing the office vacancy problem in areas with lots of vacancies (Van der Krabben and Needham 2009; Adviescommissie Plabeka 2011). The idea of this instrument is that owners of vacant buildings transfer their land and real estate in a certain area to a development company (Rensink 2011) and agree to cooperate, trying to restructure the whole area, not only their own building. After the transfer, the transformation starts and restructuring of the area begins. After demolishment and rezoning, the owners are given back new property rights, comparable to the size of the ones they initially transferred but often on a different location. The aim is that the municipality also be involved, by for example investing in the public area, as the restructuring of the abandoned offices areas is perceived as a shared problem. However, a major problem could be that by investing in one deprived area, businesses from other areas could be interested in moving to this area, as a consequence worsening the problem in another part of the city. The risk is that only existing office users relocate, and that new businesses do not come. For the city and region as a whole, such a zero sum game, or maybe even a negative sum game, would not be beneficial. The public investments might also result in market actors expecting the same in other areas: they might consider restructuring only if there also is a vast amount of public money invested. Finally, a major problem with urban land readjustment is that, even in areas with high vacancy percentages, some buildings are still occupied, making their owners reluctant to cooperate. And even owners of vacant buildings might not want to bear the risk of restructuring.

Office-for-office Instrument

Comparable with the initiative to use transferable development rights to address the problem of regional office oversupply, an 'office for office' instrument is now being discussed, whereby investors need to buy office development rights before they can start developing (Janssen-Jansen 2010a). Both TDR-like variants can be seen as use of market-based instruments to achieve more efficient land-use outcomes where regulation has failed (Spaans et al. 2011). TDR is often seen as a planning tool that reduces development potential in places that communities want to preserve, by increasing development potential in places that communities want to grow. Yet it can also be used as an instrument to convert build up area into green area or conserve heritage (Janssen-Jansen et al. 2008).

With an office-for-office instrument, a developer can develop a new office building only if he also finds a solution for the same amount of structurally obsolete office space, for example through demolition, but also transformation of the vacant building. The idea contributes both to the goal of demolishing obsolete offices or changing them to other functions and to the goal of preventing too much oversupply in the future. This idea is an example of the growing interest in ways to recoup the 'surplus value' created by planning decisions by using it to solve a part of the 'dead office' problem. It is also an example of co-production between public and private actors, although it would make public planning even more important, as it requires creating scarcity of development options (Janssen-Jansen and Salet 2009). Once again, such trade-off systems can only be implemented in the entire region at the same time, and if some of the municipalities do not want to participate in such a system, it is likely not to work. This might create a role for a regional or provincial government. In the planning of business parks with comparable vacancy problems, such a 'pay as you grow' solution has been introduced into policy ideas as well. For business parks, a restructuring fund is suggested to implement this instrument.

The Office Problem and Property Rights

The proposed ways of solving the office vacancy problem through new regulations are quite instrumental in nature, as they consist of applying incentives and disincentives. To address the office tragedy, flexible uses of property rights have been proposed and even new acts (like the Crisis- and Recovery Act) that are oriented on more collaborative solutions are more or less embedded in law. The distinction made in the introduction of this volume by Hartmann and Needham does not seem to be applicable to this lock-in situation in planning.

The proposals are about innovative ways of using law and property rights, but at the same time include aspects of more collaborative approaches. As municipalities have their own 'private' stake in office development, and have also used law and property rights to create the collective problem, it remains to be seen whether different ways of using law, regulation and exchanging property rights will result in a better situation.

Municipalities have used, and still use, public law (zoning plans) to regulate how land may be used, but for a part, they own(ed) the land themselves. The provinces, that as we have seen have been given a larger coordination task, might also have an interest in developing as much as they can, in particular as they too now have the possibility for using active land policies. Some of the provinces have experimented with regional development projects partly from a speculation perspective. Instead of balancing the development strategies of the municipalities, they increased the problem of overdevelopment, although more for housing than for commercial real estate. For the overall quality of urban and regional areas, it can be argued the provinces should improve their traditional role, instead of joining the development arena, with the risk of reinforcing the municipal failures.

The obsolete offices themselves are owned by private investors. The proposed instruments will affect the property rights of the owners of the land and the real estate. Furthermore, many of the unorthodox regulations have met criticism because they allegedly curtail procedural rights. This criticism is hardly taken into account by the relevant government bodies. Moreover, many of the owners are non-Dutch, and they might react to these regulations differently than Dutch owners.

In all the possible solutions described above, a major question is whether owners need to be compensated or not, as it is apparent that the existing property rights do matter and cannot be ignored. Many private owners of the 'dead offices', who in some of the models would lose their property, still see it as a financial asset, even though they do not use it. In some of the solutions, owners would be compensated for not continuing the non-use of his building: but should owners of 'dead offices' be rewarded for creating a problem? Or should the municipality not interfere at all? As the vacancy problems affect the overall quality of urban and regional space, it seems important for municipalities to take action.

Solving the problem will be a matter of coproduction – coproduction between private players, but also between public and private players. The question where to start remains. Many of the current initiatives do not have the support of the owners. Restructuring will then become a lengthy process. The municipality of Amsterdam could use its legal authority in a more resilient ways than before, but might want to start with the more cooperative owners in an area.

In the supply of land for offices, the government has fulfilled not so much its command and control role, but has played a more active role, even sharing the responsibility for creating the problem. More innovative and flexible ways of using law and property rights would probably result in new failures. But perhaps these new failures might be less harmful to society than the former ones.

(Fiscal) incentives can be used to encourage behaviour change of individuals and organizations. Planning is about regulations, regulations that promote socially beneficial private behaviour and sanction and limit harmful behaviour. 'Pay as you grow' planning principles might be interesting additional instruments for addressing the office oversupply and vacancy problem.

Yet, the most important way to solve the problem, and to prevent new tragedies, is that the involved players, public and private, start to realize that there is a need for a paradigm shift in planning. The fiasco of the past planning decisions, resulting in such a lock-in situation, could have been prevented if, earlier, attention had been paid to the effects of slower economic growth, partly related to known demographic change. The economic crisis has only made the problem more visible; it is not the cause. Most of the contemporary planning system, reflected in planning law, is still growth-oriented.

Planning, as we have known it, will have to change now that we have entered a period on non-growth, with limited financial resources, both private and public. Cities could face stagnation, or even contraction, while at the same time climate change challenges need to be addressed. Yet, neither public nor private players are prepared to react to these changes, and do not seem to realize their behaviour makes the lock-in situation even worse.

References

Adviescommissie Plabeka 2011. *De problemen van vandaag oplossen met de instrumenten van de toekomst*. Amsterdam: Adviescommissie Plabeka.

Allmendinger, P. and Haughton, G., 2010. Spatial planning, devolution, and new planning spaces. *Environment and Planning C: Government and Policy*, 28(5), 803–818.

BNP Paribas Real Estate 2010. *Property Report Office Market in Europe, July 2010*. Paris: BNP Paribas Real Estate.

Downs, A. 2005. Smart Growth: Why we discuss it more than we do it. *Journal of the American Planning Association*, 71(4), 367–377.

DTZ Zadelhoff 2010. *Nederland Compleet: Factsheet kantoren- en bedrijfsruimtemarkt*. Amsterdam: DTZ Zadelhoff.

Gool, P. van, Jager, P. and Weisz, R.M. 2007. *Onroerend goed als belegging*. Leiden: Stenfert Kroese.

Gordon, I. 2006. Finding Institutional Leadership for Regional Networks: the case of London and the Greater South East, in W. Salet (ed.), *European Perspectives and Randstad Holland. Synergy in Urban Networks*. Den Haag: Sdu Uitgevers, 136–160.

Haran, N. 2010. *The Power to Collaborate. How Judicious Use of Power Accelerates the Strategic Capacity of Regions in the Netherlands*. Delft: Eburon.

Hardin, G. 1968. The tragedy of the commons. *Science*, 162, 1243–1248.

Janssen- Jansen, L. and Salet, W. 2009. The Amsterdam office space tragedy. An Institutional Reflection on Balancing Office Space Development in the Amsterdam Metropolitan Region, in F. Hendriks et al. (eds), *City in Sight: Dutch Dealings with Urban Change*. Amsterdam: Amsterdam University Press, 243–260.

Janssen-Jansen, L. 2006. De tragedie van de kantoren. *Rooilijn*, 39 (8), 442–457.

Janssen-Jansen, L. 2010a. *Ruimte Ruilen [Exchanging Space]*, final report research funded by the Netherlands Organization for Scientific Research. The Hague: NICIS.

Janssen-Jansen, L. 2010b. *Ontwikkelingsbubbles en planningsdromen, Optimism Should Be in the Nature of Planners, But Over-optimism is a Dead End Street* [Dutch Planning Bubbles], essay for the annual conference of Dutch Planners and Urban Designers. Amsterdam: BNSP.

Janssen-Jansen, L., Spaans, M. and Van der Veen, M. (eds) 2008. *New Instruments in Spatial Planning: An International Perspective on Non-Financial Planning.* Amsterdam: IOS Press.

Janssen-Jansen, L.B. 2007. *Balancing Regional Developments: The Case of the North Wing Tragedy of the Offices*, Paper to be presented at the 21st AESOP Conference, 11–14 July 2007, Naples, Italy.

Janssen-Jansen, L.B. 2011. From Amsterdam to Amsterdam Metropolitan Area: A Paradigm Shift. *International Planning Studies*, 16(3), 257–272.

Krabben, E. van der and Needham, B. 2009. Land readjustment for value capturing: a new planning tool for urban redevelopment. *Town Planning Review*, 79(6), 651–672

Leo, C. 1997. City politics in an era of globalization, in M. Lauria (ed.), *Reconstructing Urban Regime Theory: Regulating Local Government in a Global Economy*. London: Sage, 77–98.

Municipality of Amsterdam 2007. *Nota hotelbeleid 2007–2010.* Amsterdam: Municipality of Amsterdam, Economic Affairs.

Municipality of Amsterdam 2008. *Toerisme in Amsterdam 2007–2008. Gasten en overnachtingen 2007 en hotels medio 2008*. Amsterdam: Municipality of Amsterdam, Department of Research and Statistics.

Municipality of Amsterdam 2011a. *Leegstand in Amsterdam: inzet op gebruik, aanpak van leegstand.* Amsterdam: Municipality of Amsterdam, department of Housing, Health and Living.

Municipality of Amsterdam 2011b. *Structuurvisie Amsterdam 2040*. Amsterdam: Municipality of Amsterdam, Department of Spatial Planning.

Municipality of Rotterdam 2011. *Convenant Aanpak Kantorenleegstand.* Rotterdam: Municipality of Rotterdam.

Ostrom, E. 1990. *Governing the Commons, the Evolution of Institutions for Collective Action*. Cambridge: Cambridge University Press.

Oudeman, P. 2010. *Werkplan Kantorenloods 2011–2014*. Amsterdam: Municipality of Amsterdam.

Peiser, R. 1990. Who plans America? Planners or developers? *Journal of the American Planning Association*, 56(4), 496–503.

Peterson, P.E. 1981. *City Limits*. Chicago, IL: University of Chicago Press.

Plabeka 2005. *Structuurschema Platform Bedrijven en Kantoren Noordvleugel, vastgesteld door de 5e Noordvleugelconferentie*. Hoofddorp: Gemeente Haarlemmermeer.

Plas, G. van der 2006. The greater Amsterdam Area?, in W. Salet (ed.), *European Perspectives and Randstad Holland. Synergy in Urban Networks*. Den Haag: Sdu Uitgevers, 85–99.

Ploeger, R. 2004. *Regulating Urban Office Provision, a Study of the Ebb and Flow of Regimes of Urbanization in Amsterdam and Frankfurt Am Main, 1945–2000*. Amsterdam: Universiteit van Amsterdam.

Rensink, A. 2011. *Samen Transformeren: Een verkenning naar de voorwaarden voor het functioneren van een herontwikkelingsmaatschappij ten aanzien van de transformatie van kantoorlocaties*. Amsterdam: Universiteit van Amsterdam.

Spaans, M., Janssen-Jansen, L. and Van der Veen, M. 2011. Market-oriented compensation instruments: lessons for Dutch urban redevelopment. *Town Planning Review*, 82/4, 425–440.

Straalen, F.M. van, Janssen-Jansen, L.B. and Van den Brink, A. 2011. *Planning for Metropolitan Landscapes: An Assessment of Dutch Regional Land Policy and Development Strategies*. Paper to be presented at the 5th PLPR conference, Edmonton, Canada, 26–28 May 2011.

Swyngedouw, E. 2005. A new urbanity? The ambiguous politics of large-scale urban development projects, in W. Salet and S. Majoor (eds), *European Cities, Amsterdam Zuidas: European Space*. Rotterdam: 010 Publishers, 61–79.

TNO 2009. *De Top 20 van Europese stedelijke regio's 1995–2007*, Randstadmonitor. Delft: TNO.

VVD, People's Party for freedom and democracy, CDA, Christian Democratic Alliance, 2010. *Freedom and Responsibility, Coalition agreement VVD-CDA 30 September 2010*. The Hague: VVD.

Chapter 11

The Fraught Relationship between Planning and Regulation: Land-use Plans and the Conflicts in Dealing with Uncertainty

Edwin Buitelaar

Planning involves institutional design. Regulations, which can be defined as legal rules that are made to structure behaviour, are important types of institutions that can be deployed for achieving aims with regard to spatial development. Planning and regulation often need each other and often go together well. Regulations are one type of resource that are employed to plan future land uses, next to for instance financial and communicative resources. And vice versa, plans such as land-use or zoning plans are often, but not always and necessarily (Neuman 1998), used to regulate land.

However, regulation is not only instrumental to planning; it can also be at odds with it. This chapter discusses the tension between both on the basis of the literature on planning and uncertainty (for example Friend and Jessop 1969; Faludi 1987; Mastop 2000) and planning and institutions (for example Bolan 1996; Gualini 2001; Alexander 2005). On the basis of this exploration, I conclude that especially strategically used (legally binding) land-use plans test the compatibility of planning and regulation.

To explore the issue further, empirical data are brought to bear on the use of land-use plans in the Netherlands. Notwithstanding the aims of the planning act from 2008, Dutch land-use plans do not plan, guide or steer spatial development, they record it and prohibit (as much as possible) deviations from these recordings. More open-ended, strategic land-use plans are impaired by rules that are aimed to fix things and minimize risks. Many of these rules originate from the field of environmental law, but also from planning law itself.

This chapter claims that there is an inherent tension between planning and regulation. Planning aims to look ahead, where regulation aims to manage what is already there. Regulation therefore involves eliminating future options and uncertainty. Planning – at least these forms that have abandoned traditional blueprint planning – tries to reduce uncertainty but also leaves, and needs to leave, room for unforeseen developments. For today's land-use plans, there are many factors, such as vested interests, environmental law and the desire to create 'certainty', which favour regulation over (strategic) planning.

Planning and Uncertainty

Planning is multifaceted and multi-interpretable. It has therefore been defined in a multitude of ways. Alexander (1992) provides an interesting overview of different perspectives on what planning is and formulates his own definition based on that review. He sees planning as:

> the deliberate social or organizational activity of developing an optimal strategy (*a plan* – EB) of future action to achieve a desired set of goals, for solving novel problems in complex contexts, and attended by the power and intention to commit resources and to act as necessary to implement the chosen strategy. (chapter 4)

Alexander takes issue with popular views that treat planning as a 'basic human activity', such as planning to go the supermarket. Although individuals do indeed plan their own daily activities, in the case of urban or spatial planning we are concerned with planning activities that seek to affect the actions of groups, organizations and government. In addition, Alexander also criticizes views, such as from John Friedmann (for example 1993), in which planning and action are fused. In line with Friedmann (1993) and Wildavsky (1973), Alexander argues that planning is not just about making plans, but also about envisioning actions that are needed to implement them. Hence, planning and action are closely related but not synonymous; the former is future-oriented while the latter is present-oriented.

The fact that planning is future-oriented implies a great role for uncertainty. Friend and Jessop (1969: 88–89) distinguish between three types of uncertainty: uncertainty with regard to the physical, social and economic environment within which decisions are made, uncertainty as to future intentions in related fields of decision-making and, finally, uncertainty as to (the development of) value judgements. The uncertainty with regard to value judgements relates to the plurality and diversity of society; in this respect, some even speak of the 'unknown society' (Van Gunsteren 1998).

The level of uncertainty varies between different types of plans. In planning theory, a distinction has been made between strategic plans and operational or project plans, which each have their own time frames and each serve their own purposes (for example Faludi and Van der Valk 1994; Mastop 2000). Strategic plans are plans that keep the future open and account for uncertainty, while project plans are blueprint-like and have a closed view of the future (see Table 11.1). Where project plans are aimed directly at the material object, strategic plans do that more indirectly via influencing the actions of subjects. In addition, strategic plans are more a step in a cyclical or an iterative (interactive) process than the end product, such as the project plan. Healey et al. show the distinct character of strategic planning by defining it as

> a social process through which a range of people in diverse institutional relations and positions come together to design plan-making processes and develop contents and strategies for the management of spatial change. This process generates not merely formal outputs in terms of policy and project proposals, but a decision framework that may influence relevant parties in their future investments and regulatory activities. (1997: 5)

Table 11.1 Project plans and strategic plans

	Project plans	Strategic plans
Object	Material	Action
Interaction	Until adoption	Continuous
Future	Closed	Open
Time	Limited to phasing	Central issue
Form	Blueprint-like	Continually updated
Effect	Well known	Frame of reference

Source: Mastop 2000

Given these features, one can actually question if project plans, in their pure form as defined here, exist on a large scale. Most planning problems are too complex, and the effects of decisions too uncertain, to treat plans as blueprints with a closed view on the future. Some form of strategic action is often required. Dealing with limited knowledge and uncertainty also implies flexibility, we have moved beyond the times of blueprint or end-state plans (Taylor 1998). Flexibility also means leaving options open instead of prescribing directions in detail.

Planning as Institutional Design

Much of what planning does is institutional design (Bolan 1996; Alexander 2005). Some even go as far as to say that planning *is* institutional design (Innes 1995). Institutions help guiding future developments and future interaction. Adopting a land-use or zoning plan and drawing a spatial vision for a particular territory are indeed acts of institutional design. But also the active development of locations by developers or local authorities can be seen as maintaining or deviating from institutions with regard to land development, and hence as (re)designing institutions.

Regulations can be seen as institutions in as far as they have become socially accepted and embedded. Earlier, it has been argued that the extent to which rules are institutionalized is indicated by the actual use of them (Buitelaar, Galle and Sorel 2011). This implies that not all regulations within planning law are institutionalized, or at least not at every level. For instance, the Dutch planning act contains the instrument of the 'conservation ordinance' (*beheersverordening*). Although its introduction in the act is the result of an institutionalization process at the central political and administrative level, locally it has not become institutionalized – for various reasons – since it is seldom used.

Institutionalized regulations are key instruments for implementing planning policies. Zoning rules, subdivision rules, rules for septic systems, minimum lot-size requirements, density rules, infrastructure requirements and so on are often seen as essential means to shape urban development in a publicly desired direction. However, at the same time, they have the potential to inhibit planning. This is related to the sometimes contradictory aims of planning and regulation. Where planning often seeks to make things happen, regulations are more likely to prevent things from happening.

This has to do with the *raison d'être* of regulation. There is a clear link between the emergence of regulations – in particular public law rules – on the one hand and the growing complexity of modern society, and the risks associated to that, on the other. The emergence of the 'risk society' (Beck 1989) and the 'regulatory state' (Majone 1994) seem to go hand in hand.

> The growth of administrative regulation in Europe owes much to (these) newly articulated perceptions of a mismatch between existing institutional capacities and the growing complexity of policy problems: policing financial markets in an increasingly interdependent world economy; controlling the risks of new products and new technologies; protecting the health and economic interests of consumers without impeding the free flow of goods, services and people across national boundaries; reducing environmental pollution. (Majone 1994: 85 – brackets by the author)

Regulations take the place of more direct forms of government intervention such as central planning and public ownership. They have the ability to create some certainty in highly complex and individualized societies.

Not only do rules have the ability to reduce risks for society, they can do the same for policy-makers and politicians. Weaver (1986) argues that policy-makers are inherently blame avoiding, rather than credit claiming because 'voters are more sensitive to what has been done *to* them than to what has been done *for* them' (Weaver 1986: 373). This inclination has become stronger with an increase of *mediatized politics* (Hajer 2009). Nowadays, political mistakes are illuminated, inflated and disseminated under large numbers of people in a short period of time. The chances of personal mistakes are reduced through regulations since these standardize, eliminate or overtake discretionary policy decisions. This is in line with Weber's bureaucratic ideal (Weber 1922).

So, one of the main aims of regulations is to minimize or eliminate risks, both for society and government. They are aimed at eliminating future options, where (strategic) planning is also about leaving options open. Legally binding land-use or zoning plans are plans where planning and regulation come together explicitly. Many would categorize this type of plan as project plans. Non-binding, indicative plans such as the Dutch structure vision, the English development plan or the European Spatial Development Perspective (ESDP), usually qualify as strategic plans (e.g. Needham 2000, footnote 1). However, this distinction is somewhat simplistic and does insufficient justice to the complex and hybrid reality, in which plans combine features of both. Land-use plans, for instance, can also and sometimes do fulfil a more strategic function. I distinguish between three different functions of the land-use plan with regard to managing land uses. First, it can be used to conserve or preserve existing land uses. Plans that do that primarily are what I later call *conservation plans*. Second, plans can be used to facilitate development initiatives that deviate from the existing regime of land-use planning. In the US, this is called rezoning. Since these plans follow on development initiatives, I call them *development-led plans*. Both these types of plans, or functions of the plan, share most of the features of the project plan, as defined earlier (Table 11.1). The third function is to guide future land uses in a direction that government wants, and deviates from the existing land use, while there is no developer or development initiative yet. One can think of a plan that makes room for housing on a monofunctional business estate so as to create a more mixed and vibrant built environment. Plans that have this future orientation as their dominant function are called *steering plans*.[1] These plans are more strategic than the other two categories since they leave the future more open, not knowing exactly what lies ahead.

Strategic land-use plans need, in order to be strategic, more broad regulations rather than detailed. Otherwise, they would be too restrictive, making the acceptance of unknown, but potentially desired, developments more difficult. In the next section, I will also demonstrate empirically that although (strategic) planning often needs regulation to implement policies, integrating both in more strategic land-use plan seems a contradiction.

Dutch Land-use Plans: Between Planning and Regulation[2]

In 2008, the Netherlands got a new spatial planning act, the Wro, which replaced the previous act: the WRO. One of the aims of the new act was to (re)instate[3] the land-use plan (*bestemmingsplan*) in a central position in development control.

1 In Buitelaar, Galle and Sorel (2011) we use the slightly awkward term *plan-led plans* instead. 'Steering' seems to capture better the nature of these plans as ways of guiding future development instead of confirming them *ex post*. 'Steering' refers to the intention of the plan; whether these plans are actually successful in doing that is a different matter.

2 This section is largely derived from Buitelaar, Galle and Sorel (2011).

3 Whether the *bestemmingsplan* ever had a central role is open for discussion.

A practice had emerged in which exemptions from the land-use plan, instead of the plan itself, were used to facilitate development (Buitelaar and Sorel 2010). In addition, government wanted the land-use plan to fulfil a steering or strategic function, in the sense that it should provide guidance for future developments, instead of only being the *ex post* formalization of development initiatives. In this section, I describe how this worked out in the first year and a half after the introduction of the act. I make use of both quantitative data and qualitative information from interviews with twelve municipalities (see for more information PBL 2010; Buitelaar, Galle and Sorel 2011). At the moment of writing, the law is being changed again; the effect of that is not taken into consideration. Therefore, the empirical report is written in the past tense.

For the land-use plan to take on its central role various conditions were added to the new act. One of them stipulated that the land-use plan was mandatory for the entire municipal territory, not just for land outside the built-up area, which was the case under the old act. Another condition was the sanction of not allowing municipalities to charge building fees, when a land-use plan had not been revised within ten years. In addition, the number of exemptions and the possibilities to apply them were reduced. Article 19.1 (and 19.2) exemptions were replaced by the project decision (*projectbesluit*). The addition of the project decision was urged by municipalities and property developers, as they saw a need for a simpler and quicker instrument in the daily practice of planning. The project decision was seen as the first stage of granting the building permit, while detailed elaboration of requirements could wait till the revision of the land-use plan. However, the circumstances under which the project decision can be used were less attractive. The project decision had to be succeeded by a revised land-use plan within one year. If that did not occur, then again local authorities were not allowed to charge building fees.[4] In addition, the procedure of the project decision provided two possibilities of appeal, while appeal against the land-use plan could take place in only one instance. Figure 11.1 shows the result of this in terms of the difference in the balance between the land-use plan, and its alternatives under the old and the new act (WRO is the old planning act, Wro the new).

As argued before, under the old act, exemptions were the dominant regulatory mode for the granting of building permits. Under the new act this is reversed: the land-use plan was used several times more than the project procedure.[5]

In the interviews we were given different reasons for the dominance of the land-use plan. First, as might be expected, municipalities see the sanction of not charging building fees as a serious disincentive. Avoiding that sanction by putting a land-use plan on public display within a year after the project decision was not

4 On 31 March 2010 this obligation was abolished as a part of the Crisis and Recovery Act, an act introduced to streamline decision-making on spatial projects as a means to economic recovery.

5 The increase of exemptions in 2008, just before the introduction of the act, indicates that change was going to happen, a change that many did not prefer or were cautious of.

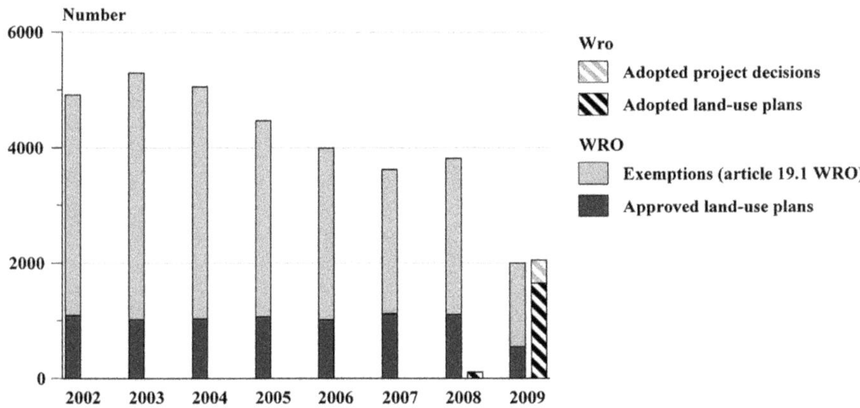

Source: PBL 2010.

Figure 11.1 Land-use plans, project decisions and exemptions under WRO and Wro

regarded as a solution, since that would mean running the same procedure twice. In addition, the possibility of two appeal moments instead of one made the project decision less attractive. A last, though minor, disadvantage of the project decision compared to the land-use plan was that it did not provide the same options for including flexibility. For instance, a land-use plan could include the option to postpone further specifications to a later stage, an option that was used in parts of virtually all (96 per cent) land-use plans (PBL 2010).

There were however also some advantages of the project decision over the land-use plan that made application interesting for some municipalities. First, the project decision could be delegated by the elected council to the local administration, while the land-use plan could not. This led to a difference of seven weeks (12 weeks as opposed to 19 weeks) in average procedure time between the public display of the draft land-use plan and the adoption thereafter. Second, some local authorities considered the project decision as a slightly 'lighter' instrument than the land-use plan and therefore used it indeed as a means to organise decision-making into phases, by first deciding about the development as such, while deciding later on the detailed regulations.[6] Third, municipalities that were in the process of drafting a land-use plan for a larger area sometimes used separate project decisions to fast-track developments that fitted in the forthcoming overall land-use plan. Since a

6 Whether this is entirely correct in legal terms is questionable. In a recent court case at the council of state (*Raad van State*, case 2010 04647/1/H1) on a project decision it was judged that it provided too little detail. The judge argued that the purpose of a project decision is to facilitate a concrete project, not future potential developments. So *de facto*, a project decision should contain the same elements and level of detail as the land-use plan.

new plan was being made, the condition to revise the current land-use plan within a year could be met easily.[7]

Land-use Plans and their Steering Function

Statistically, the land-use plan had become the key document against which building permits are tested. Therefore, one could argue that it had (re)gained a central role. However, the question remains if it also fulfilled its intended strategic or steering function.

In order to answer that question, we investigated 209 adopted digital land-use plans (PBL 2010), 12 per cent of the total number of adopted land-use plans in the research period. These have been analysed according to the motive behind the plan, the initiator, and the plan's relation with the existing[8] land use. We distinguished between three categories of plans (see Figure 11.2).[9] The development-led plan is the first. As said earlier, these are plans that are made or revised as a response to a concrete development initiative, from either a public or a private actor, which does not fit within the existing land-use plan. Usually the building permit is granted right after the adoption of the plan. 53 per cent of the plans lie in that category. Whether these are 'strategies of future action', as Alexander (1992) defines 'planning' is open for discussion. The second category contains conservation plans. These plans are made to maintain the current land-use pattern and property rights. They are therefore often very detailed and precise (see Figure 11.2). Many local authorities are updating their obsolete land-use plans, using this type. As a result, 35 per cent of the land-use plans are conservation plans. Hence, almost 90 per cent of the land-use plans adhere to most of the features of the project plan, as defined earlier (Mastop 2000). However, it needs to be noted that these plans too, have strategic elements and allow for flexibility to keep future options open. Virtually every land-use plan in the Netherlands includes conditions that prescribe the circumstances under which deviation from or elaboration of the land-use plan is allowed (PBL 2010). In all these cases, consent from local government is required.

No more than 8 per cent of the plans are steering or strategic. These are plans that do not yet have an initiative or initiator and that allow for, and guide, unknown developments by unknown developers. In line with the discussion earlier these can be called strategic land-use plans.

Rather than planning and guiding development, the land-use plan primarily recorded development initiatives or maintained existing land uses. The majority of the plans were development-led. Most of them were relatively small plans, also called 'postage stamp plans'. On average, the territory that development-led plans

7 This practice might become more common, now that the obligation to follow a project decision with a land-use plan within a year has been abandoned with the introduction of the Crisis and Recovery Act.

8 'Existing' at the moment of the adoption of the plan.

9 There is also a fourth category that contains a mixture of the three other categories.

Source: PBL 2010

Figure 11.2 Examples of bestemmingsplannen

covered was around ten times smaller than that of the conservation plan (PBL 2010). Many of these plans had taken over the role that the permitted exemptions under the old act (section 19.1 and 19.2 WRO) used to play. It raises the question whether the majority of the land-use plans were *plans* and whether a planning system can be called 'plan-led' just based on the fact that building permits are tested against *de jure* plans. *De facto* the majority of land-use plans showed more similarity with a *contract* in which informal agreements are formalized *ex post*.

Factors Responsible

In an earlier paper, we made the case that steering or strategic land-use plans are financially unattractive for local authorities. For it is possible to pass on the costs of making the plan to another party only when there is already a concrete development initiative. The costs of plans without an initiator have to be covered by the local authority. This line of reasoning fits within in the institutionalized planning practice – the planning culture – in which plans are used pragmatically and strategically (Buitelaar and Sorel 2010).

However, interviewees and additional literature (Van der Velde 2009) indicated that there is an additional set of regulatory factors that inhibit the adoption of steering plans. A strategic land-use plan requires a more abstract and open formulation of land uses and conditions, since it seeks to accommodate development initiatives that are not yet known. Not only is it considered difficult to prescribe or allow for future developments without creating an inflexible regime for these developments, even when this is done in a rather broad way, there are also various formal regulations that hamper the adoption of this type of plan and consequently require more detailed plans.

One set of conditions is formed by environmental laws that have become institutionalized at the local level over the last 20 to 30 years, such as those related to soil contamination, noise nuisance, air quality, flora and fauna protection and external safety. For instance, in order for a plan to comply with noise nuisance rules there must be some indication of the noise level of the developments that are to take place. Open-ended plans provide too little certainty about potential hazards. Court cases also show that the court has difficulty with plans that provide too little certainty.

However, it is not only the friction between planning law and other fields of law that deters the making of strategic plans, but also the tensions within the planning act. First, the act demands from local authorities that they demonstrate that implementation of the plan is financially feasible. This implies some precision in the designation of land uses, and a connection to concrete development initiatives. Second, the act requires from local authorities that they recover all public costs related to the development of the plan area.[10] These include the costs of servicing the land, infrastructure and plan-making. But when land uses are not exactly clear, the costs are difficult to calculate and to attribute. Steering plans and cost recovery, as defined in the current act, do not go well together. The same applies to planning compensation. If land-use designations are too broadly defined, there is a greater chance that adjacent landowners successfully claim that the plan decreases the value of their property – the impact of the allowed development is greater – and that they are therefore eligible for planning compensation. This can be a large financial burden for local authorities.

The Future of the Land-use Plan

This chapter shows that planning, which has to take account of uncertainty about the future, has a problematic relation with regulation. Where regulation seeks to eliminate future options so as to create certainty, and avoid risk and blame, strategic planning implies dealing with uncertainty by making plans flexible and adaptive.

What is the implication of that in the Netherlands, and elsewhere? Can legally binding land-use plans be used to plan strategically? Based on the empirical work reported here, I am inclined to say 'no'. Development-led and conservation plans give guidance to future developments only in the sense they prohibit almost everything else than the current use and concretely proposed use respectively.

Is that problematic? Maybe the idea of a steering or strategic land-use plan must be abandoned. Visionary, indicative, strategic spatial plans (in the Netherlands called structure visions) are suitable for setting long-range aims for an uncertain future. These plans are not aimed at prescribing the material object (for example land use) but at structuring and informing decision-making processes about land use (Mastop 2000). The plan is not the end of the process, but a step in an ongoing sequence of events.

Land-use plans could (or should) 'only' be used as (conservation) ordinances – which are local, more generic laws instead of location-specific plans – that regulate land uses and provide legal certainty. The question is: how? If this is done

10 With the total value of the serviced land as the ceiling for cost recovery.

in great detail, the chances are great that development initiatives that are favoured by the local planning authority require a revision of, or a deviation from, the existing land-use plan. This leads not only to bureaucracy and transaction costs, but in the long run also to less material legal certainty (Buitelaar and Sorel 2010; Moroni 2007).

These deviations from the land-use plan or ordinance can be supported by something like the earlier mentioned project decisions, which fulfil the role that the development-led land-use plans fulfil in the Netherlands. When functioning well, decisions to deviate from the existing land-use regime should be informed by a non-binding strategic spatial plan. If the development proposal does not fit within that plan it should not be rejected straight away. There can be good reasons to facilitate the proposal. The strategic plan helps to make that reasoning explicit. In this model, planning and regulation are clearly related sequentially, but not integrated in one land-use plan.

References

Alexander, E.R. 1992. *Approaches to Planning: Introducing Current Planning Theories, Concepts and Issues*. Luxembourg: Gordon and Breach Publishers.

Alexander E.R. 2005. Institutional transformation and planning: from institutionalization theory to institutional design. *Planning Theory*, 4(3), 209–223.

Beck, U. 1992. *Risk Society: Towards a New Modernity*. London: Sage Publications.

Bolan, R.S. 1996. Planning and institutional design, in S.J. Mandelbaum, L. Mazza and R.W. Burchell (eds), *Explorations in Planning Theory*. Center for Urban Policy Research, Rutgers, University of New Brunswick, New Brunswick, NJ, 497–514.

Buitelaar E. and Sorel, N. 2010. Between the quest for control and the rule of law: legal certainty in the Dutch planning system. *Land Use Policy*, 27(3), 983–989.

Buitelaar, E., Galle, M. and Sorel, N. 2011. Plan-led planning systems in development-led practices: an empirical analysis into the (lack of) institutionalisation of planning law. *Environment and Planning A*, 43(4), 928–941.

Faludi, A. 1987. *A Decision-Centred View of Environmental Planning*. Oxford: Pergamon Press.

Faludi, A. and van der Valk, A. 1994. *Rule and Order: Dutch Planning Doctrine in the Twentieth Century*. Dordrecht: Kluwer.

Friedmann, J. (1993), Toward a non-Euclidian mode of planning. *Journal of the American Planning Association*, 59(4): 482–485.

Friend, J.K. and Jessop, W.N. 1969. *Local Government and Strategic Choice: An Operational Research Approach to the Process of Public Planning*. London: Tavistock Publications.

Gualini, E. 2001. *Planning and the Intelligence of Institutions: Interactive Approaches to Territorial Policy-Making between Institutional Design and Institution-Building.* Aldershot: Ashgate.

Hajer, M.A. 2009. *Authoritative Governance: Policy-Making in the Age of Mediatization.* Oxford: Oxford University Press.

Healey, P., Khakee, A., Motte, A. and Needham, B. (eds) 1997. *Making Strategic Spatial Plans: Innovation in Europe*. London: UCL Press.

Hood et al. 1999. Where risk society meets the regulatory state: exploring variations in risk regulation regimes. *Risk Management*, 1(1), 21–34.

Innes, J. 1995.Planning theory's emerging paradigm: communicative action and interactive Practice. *Journal of Planning Education and Research*, 14(3), 183–189.

Majone, G. 1994. The rise of the regulatory state in Europe. *Western European Politics*, 17(3), 77–101.

Mastop, H.J.M. 2000. The performance principle in strategic spatial planning, in W. Salet and A. Faludi (eds), *The Revival of Strategic Spatial Planning*. Amsterdam: KNAW, 143–155.

Moroni, S. 2007. Planning, liberty and the rule of law. *Planning Theory*, 6(2), 146–163.

Needham, D.B. 2000. Making strategic spatial plans: a situational methodology, in W. Salet and A. Faludi (eds), *The Revival of Strategic Spatial Planning*. Amsterdam: KNAW: 79–90.

Neuman, M. 1998. Does Planning Need the Plan? *Journal of the American Planning Association*, 64(2), 208–220.

PBL 2010. *Ex-durante evaluatie Wro: eerste resultaten*. The Hague: Planbureau voor de Leefomgeving.

Taylor, N. 1998. *Urban Planning Theory Since 1945*. London: Sage publications.

van der Velde, J. 2009. *Met globale bestemmingsplannen meer mogelijkheden.* The Hague: Ministerie van VROM.

Van Gunsteren, H.R. 1998. *A Theory of Citizenship: Organising Plurality in Contemporary Democracies*. Oxford: Westview Press.

Weaver, K. 1987. The politics of blame avoidance. *Journal of Public Policy*, 6(4), 371–398.

Weber, M. 1922. *Economy and Society*. University of California Press.

Wildavsky, A. 1973. If planning is everything, maybe it's nothing. *Policy Sciences,* 4(2), 127–153.

Conclusion
Planning by Law and Property Rights Reconsidered

Barrie Needham and Thomas Hartmann

When land-use planning, or spatial planning, is practised, planning law and property rights interact, and the intention is an outcome that a government 'planning agency' desires. This practice is very common in many countries of the world, but there has been surprisingly little theoretical attention given to it. That gave the occasion for writing this book. We have investigated this activity of 'planning by law and property rights' by first setting out what we understand it to be (Introduction) and then by analysing – theoretically and by game simulation – the relationships between (public) planning law and (private) property rights (Chapters 1 and 2). Then we have placed planning law and property rights within their wider institutional setting at three different scales: when a country undergoes enormous social and political change (Chapter 3), when there is fundamental political debate about the extent of the influence of the state within a country (Chapter 4), and when a country changes its legislation in response to a policy of the European Union (Chapter 5). Global environmental change poses huge challenges for spatial and environmental policies, and the following three chapters were about how planning by law and property rights can respond to those challenges: by adaptive planning (Chapter 6), by adaptable property rights (Chapter 7), and by public policies at the appropriate geographical level (Chapter 8). Then the book returned to the issue emphasized in the Introduction: planning by law and property rights can determine a local regime of property rights which might turn out to be inappropriate but difficult to change. When that has happened, how can that regime be changed (Chapter 9)? Can planning agencies make such undesirable lock-ins less likely to happen by reducing market uncertainty (Chapter 10)? What type of planning should be practised and what type of plans made in order to reduce the probability of creating undesirable lock-ins (Chapter 11)? In this final chapter, we draw conclusions from that investigation, conclusions related to the general issues raised in the Introduction.

The Nature of Planning by Law and Property Rights

The exploration of the previous chapters enables us to define better what we mean by 'planning by law and property rights'. This way of spatial planning searches for modes of influencing land use by applying, and if necessary adapting, the planning law, alone or in active combination with property rights. Then, land use is influenced by regulation – applying the law, an emphasis on state activity – and/or by stimulating markets in property rights.

The regulation can take two forms. One is to state the conditions which land use must satisfy. For example, if one wants to realize a commercial use in a residential zone, it may not be closer than a certain distance to a kindergarten. The other is to state the desired result. For example, if one wants to realize a commercial use in a residential zone, that land use may not interfere with a kindergarten. These are two different types of zoning, the first would be what US Americans would call 'Euclidian zoning', the latter is 'performance zoning' (Elliott 2008: 9–38). In legal terms, this is the distinction between condition-based and performance-based regulation (see Chapter 5).

That is regulation of the way in which property rights in land and buildings are used. An alternative way of influencing land use is to determine property rights in a particular way and let them be defended or exchanged in markets. Prominent examples are emission rights or transferable development rights. Needham discusses in chapter 1 how planning by property rights might be an alternative to regulation. Ostrom has shown that it is possible to govern the commons – to avoid the tragedy – by deconstructing the bundle of private property rights into five single rights (access, withdrawal, management, exclusion, and alienation, see also Ostrom 2009: 28).

Law and Property Rights: Inevitable and Desirable

After discussions in planning theory about planning in the aftermath of post-positivism, post-industrialism, or post-Fordism (see e.g. Allmendinger 2002, or Baum 1977), we seem to have reached a period of the post-post-isms, which is guiding spatial planning into a direction that neglects law and property rights and considers it to be an old-style top-down approach. This book counters the move in that direction.

Planning affects the allocation and distribution of land use (e.g. Davy 2005; Needham 2006). The essence of this statement is that planning is a way of formalizing the social construction of the burdens and benefits of landownership. This formulation makes it clear that even the deliberate decision not to interfere in a certain arrangement of rights is a planning decision. Public authorities are empowered – within constitutional boundaries – to change property rights and impose or relax regulations. So they always have options, between doing nothing and doing something, and if they choose to do something, what should that be? There is no alternative to planning by law and property rights. The only options are: planning by law and property rights alone or in combination with other approaches, such as collaborative planning. Even the libertarian perspective of reducing planning to a minimum, and 'let people themselves decide how to use their own land and property' (Needham 2006: 3), requires a decision about planning, law, and property rights, namely regarding the definition and the initial assignment of the rights – as Ronald Coase has discussed in his famous article 'The problem of social cost' (Coase 1960).

A corollary is that spatial planning is embedded in the wider institutional context of the country, in any case in those countries which claim to have a rule of law and a coherent legal system. The planning law, property rights, and the way in which they interact should fit within that wider system. That imposes constraints on the way in which planning law is applied and on possible changes to it. A second corollary is that knowing how to carry out such planning should be one of the key skills of spatial planners. And the syllogism that results from those two statements is that planners should know the law, and the system in which it is embedded. 'If a policeman must know the Constitution, then why not a planner?' – William J. Brennan Jr., Supreme Court Justice (1906–1997).

Planning by law and property rights is not only inevitable, for the above reasons, it is also desirable. One reason is that it can provide a stable framework for spatial developments, and this enables public and private stakeholders to make their decisions in a robust and reliable context. A zone in a zoning plan allows an investor to build – let us say – housing, thus investing money based on estimated future profits. And this investment is secure because, in a sound system of planning law and property rights, the right created by the zoning plan may not be taken away, at least not without appropriate compensation. If the adjacent area is zoned 'not-for-building', this can affect the enjoyment from housing constructed in the building zone, and therefore its value: the occupier and the developer want to be secure in this too. But note that this value of planning law and property rights is enjoyed only if the plans are not changed frequently and only if the plan is applied consistently to guide development.

The other reason why planning by law and property rights is desirable is that it can satisfy the principles of good governance. Spatial planning of whatever sort is an exercise of the public administration and should, therefore, satisfy democratic principles which the citizen demands of that. These include openness, objectivity, rationality, fairness (and see Needham 2007: 112 et seq.). Planning by law and property rights can satisfy these principles because it requires the application of laws which include guarantees for protecting the rights of the citizen. Here also the caveat applies: this value of planning by law and property rights is enjoyed only if the planning laws are applied correctly.

Wicked Problems and Lock-in Situations

In order to carry out purposeful planning by law and property rights, it is necessary to predict the intended and unintended effects. Even the decision to do nothing is based on a prediction of the effects of that! How do people react to changes in property rights and in public regulations which affect the exercise of those rights?

Who reacts and in what ways? In the short run and in the long run? What are the effects on land use in the first instance? And what are the consequent social effects, economic effects, and environmental effects? How good is that knowledge? Does it allow the effects to be predicted reliably? To what extent is such prediction possible?

The effects of small planning measures can usually be predicted fairly reliably. An example is that people may not build in their back gardens. But ambitious spatial planning is concerned with what this book has called 'wicked situations' – following the terminology of Rittel and Webber (1973). What makes spatial planning on the one hand so exciting and, on the other hand so wicked, is that spatial planners are in a dilemma: they cannot fully decide about the distribution and allocation of land uses, but according to most planning laws, this is precisely their task. Moreover, the actual allocation of land uses and the distribution of its costs and benefits depend on decisions of many different stakeholders. For this reason, 'the planners should not be preoccupied with urban forms, but instead focus on the social processes operating in cities' (Mattila 2002: 131).

How can planners better predict and explain how owners of property rights respond to changes (or no change) in law and property rights? Hillier regards planning as an 'experiment with unthought possibilities' (Hillier 2010: 14). Policy experiments are essential for finding successful policies, states Bertolini (2010: 86) in his contribution to the book *A Planner's Encounter with Complexity*. But to what extent is it possible to experiment with planning law and property rights? Changes in the legal system tend are difficult to introduce. Changes in the law often do not have the desired effect.

This book is not primarily about how to predict the effects of spatial planning, but the difficulty in doing that lies at the core of the book. For the central question that we pose is: given the fact that it is so often so difficult to predict the effects of spatial planning, especially in the longer term, and given that this can result in plans which produce undesirable lock-ins, what type of planning should be practised?

The great strength of planning by law and property rights is that it creates stable conditions for owners and users of land, for it gives them property rights which can be defended in law. That strength is also a weakness, for it might turn out to be desirable and sensible not to apply a particular plan or planning policy after it has been formally adopted. The reasons we have just discussed: that the choice of planning measures depends on predictions of their effects, and often those predictions cannot be made reliably. The physical environment might have changed permanently; the wider institutional context might have been changed; the market – what is demanded, what is supplied elsewhere, the prices – might have changed.

Under such circumstances, it can be very sensible to depart from or to change the plan. But that might reduce the value of those property rights: and as they can be defended in law, changing them can be very difficult and/or expensive. As a result, planning by law and property rights can become inflexible and produce lock-in situations.

Unlocking a Lock in

Lock-in situations can be overcome only with high transaction costs, as Arthur showed in his research on competing technologies (Arthur 1989). Suppose that a lock-in has arisen. How can it be 'unlocked'? Unlocking a lock-in requires an 'optimal perturbation' (Thompson 1997: 214) to disturb a rather robust situation. An optimal perturbation can take various forms. The optimal perturbation in the case of the Amsterdam office space tragedy (see Chapter 10) would probably be financial – namely demolishing office buildings or reconstructing them so that they can be used differently. In the case of the German water management (see Chapter 5), the lock-in is institutional – so the optimal perturbation would be something like an institutional reform. In case of climate change, the 'lock' into a certain undesired property rights regime – as Tarlock describes it (see Chapter 7) – might require a different perturbation, such as an adaptation of property rights. Once a situation in spatial planning is locked-in, unlocking it will cost a lot of money; and in spatial planning this is usually at the expense of the public sector.

Preventing and Managing Lock-ins

So, if unlocking is to be avoided, the challenge is to prevent lock-ins emerging. There are two possible strategies for doing this: anticipation, and inflexibility reduction.

Anticipation

If we could anticipate better how institutions, the environment, and markets will react to a plan, it would reduce the inherent uncertainties in planning and so reduce the probability that lock-ins arise.

That is, however, easier said than done. Chapters 9 and 10 show that even when municipalities make forecasts on which to base decisions about their own, substantial, capital investments, they can get those forecasts seriously wrong. Why should private investors believe that municipalities can forecast better than they do? In particular, when we are dealing with wicked problems, which cannot be finally defined, have no stopping rule, and are always unique (Rittel and Webber 1973: 161–167), forecasting is always very difficult.

Anticipation is, however, not the same as forecasting. It is a method which says: if we do that, how might others react to it? What can we do now, what can we build into the plan or the planning, so that the desired reactions to the plan are encouraged and the undesired made less probable or less damaging? In that way, anticipation can reduce uncertainties.

One way of doing that is – as proposed by Needham and Geuting (Chapters 1 and 2) – by making use of thought experiments respectively game simulations.

Another way is to incorporate other parties in the anticipating process. For example, the ambitions for spatial planning could be scaled down, by wanting to regulate fewer aspects of the physical environment, and allowing more scope for private persons and firms. Fewer detailed plans, more plans of a general nature. If this were to be chosen, it would mean accepting that the plans provide less certainty to private investors, developers, and users. The disadvantages of that cannot be known without more knowledge of the kind and amount of certainty about the physical environment that people want. What kind of certainty do private persons and firms actually value, what types of uncertainty do they want to avoid? Who, within a given plan area, wants which type of certainty and who would benefit from a particular kind of uncertainty? A variant of 'less planning' is to create property rights of such a nature that the ensuing market in those rights will create the situation desired by the planning agency: but only if 'the market' wants to. If circumstances change, whereby 'the market' does not want to do that, then a planned but ultimately undesired situation will not arise. Another way of incorporating the market (in a wider sense) is to involve private persons and firms more deeply when producing the plans. This could result in the plans meeting better the future situation for which they are now made. However, the idea that, by incorporating the market by means of joint ventures between public and private investors, plans will be made which have less chance of being wrong is not supported by the cases given in Chapters 9 and 10 in this book.

Such closer working with other parties is what is propagated by collaborative planning schemes. Planning by law and property rights may not neglect such schemes, but should make a good use of them, in order to achieve better planning by law and property rights (not replacing it!). Collaborative planning is indeed not the opposite of planning by law and property rights: with the introduction of the idea of collaborative planning, Healey wanted to

> expand imaginative range of those designing new process forms beyond the well-known possibilities of manipulative politics, the rational-technical process, top-down command and control practices and bureaucratic rule-governed behaviour. (2003: 107)

Collaborative planning, in its original form, is thus not an alternative to planning by law and property rights, but an extension of it. Moreover, we have seen that it cannot replace planning by law and property rights, which must always be present in one way or another, if only to make legally binding the agreements reached under collaborative planning. Collaborative planning cannot be used instead of planning by law and property rights, but it might be an appropriate extension to reduce uncertainties.

Under what circumstances would collaborative planning be useful in coping with wicked situations? Particularly in situations where the implementation of certain planning goals depends on stakeholders outside the jurisdiction of the spatial planner, such as for transnational planning issues like flood protection, but also for collaboration between market actors and governmental institutions. However, we have seen, particularly in part four of the book (see Chapters 10 and 11) that collaboration (e.g. between municipalities, between public and private stakeholders) can also lead to lock-in situations, especially if one of the reasons that private parties want to collaborate is that public parties take some of the financial risk, with the result that private parties become less cautious.

Inflexibility Reduction

The alternative to anticipation is the reduction of inflexibility in planning. Then, lock-in situations can be avoided. Thompson states that, for complex systems – and wicked problems result from complexity (Rittel and Webber 1973) – we need to 'switch from anticipation to inflexibility reduction, then we will have the basis for the sort of intervention we need' (Thompson 1997: 214).

Following this argument, the key question is: how to make spatial plans less inflexible without abolishing the desired stability and security that law and property rights create?

One way is by creating property rights which are adaptable to changed (environmental) conditions, as Tarlock is proposing (Chapter 7). This would at least redistribute the burdens of environmental change so that they fall not on the state, but on the owners of the property rights. We need to consider such proposals in the light of the challenges of spatial planning. In Chapters 1 and 2 we discussed methods for experimenting with such ideas.

'The challenge lies not in the *planning for adaptation*, but in *adaptive planning*', concludes Tennekes (Chapter 6). Facing rapid environmental changes, planning should not try to find once-and-for-all solutions, but practise adaptive planning. This means creating an institutional design that frames agenda setting, gives incentives to market parties, allocates responsibilities and risks in a way that does not aim to find robust and long-term planning solutions for future challenges – the aims of the positivistic (Allmendinger 2002) and the rationalistic (Etzioni 1967) views on planning – but permanently to take into account the newest insights into the planning challenges.

Inflexibility can be reduced also by making small scale 'project plans' (see Chapter 11), which take account of the wider geographical situation by fitting into a spatial plan which is only indicative, not binding. Such a project plan is made at the last moment, just before implementation. However, although that reduces the chance that a plan becomes out of date before it is implemented, it does not prevent a situation arising later where someone wants to change 'the project' but cannot do so because of the (binding) project plan.

Another way could be to break out of institutional entrenchments. It is essential to know the institutional setting. Both condition-based legislation and performance-based legislation have advantages and disadvantages. Condition-based regulation provides more institutional stability and legal security; whereas performance-based regulation is more adaptive and more open to new solutions.

Chapter 5, on German water management, shows how ignoring (or at least not sufficiently considering) institutional settings can lead to unexpected difficulties. And the paradigm shift proposed there seems to be very sensible.

Don't Reject and Replace but Reconsider

This book has put forward various ways of avoiding getting unwittingly 'locked in' to an undesirable situation, by a particular application of law and property rights in spatial planning. It has shown also that collaborative planning can be a useful extension of planning by law and property rights, but is no solution to avoiding lock-in situations in the face of wicked problems. For that reason, rejecting and replacing law and property rights is not an option.

Moreover, we must never forget that spatial planning is more than a technocratic exercise, where effectiveness is the only criterion of success. For example, it might be decided to plan by planning law regulations rather than by joint ventures, even though that latter might be more effective in some circumstances, just because the former is more transparent and accountable than the latter.

We close by repeating the main message of this book. Planning by law and property rights does not cope well with uncertainties and wicked situations. This is not a reason for rejecting it and replacing it with another approach to planning. For planning by law and property rights is inevitable, and it is desirable too. Planning by law and property rights needs to be reconsidered, so that it can be made to work better.

References

Allmendinger, P. 2002. Towards a post-positivist typology of planning theory. *Planning Theory*, 1(1), 77–99.

Arthur, W.B. 1989. Competing technologies increasing returns and lock-in by historical events. *The Economic Journal*, (99), 116–131.

Baum, H.S. 1977. Toward a post-industrial planning theory. *Policy Sciences*, 8(4), 401–421.

Bertolini, L. 2010. Complex systems, evolutionary planning?, in G. de Roo and E.A. Silva (eds), *A Planner's Encounter with Complexity*. Farnham: Ashgate, 81–98.

Coase, R. 1960. The problem of social cost. *Journal of Law and Economics*, (3), 1–44.

Davy, B. 2005. Bodenpolitik, in by E.-H. Ritter (ed.), *Handwörterbuch der Raumordnung* (4th edn. Hannover: ARL, 117–130.

Elliott, D.L. 2008. *A Better Way to Zone: Ten Principles to Create More Liveable Cities*. Washington, DC: Island Press.

Ellis, R.J. and Thompson, M. (eds) 1997. *Culture Matters: Essays in Honor of Aaron Wildavsky*. Boulder, CO: Westview Press.

Etzioni, A. 1967. Mixed scanning –. A. 'third' approach to decision-making. *Public Administration Review*, 27(5), 385–92.

Healey, P. 2003. Collaborative planning in perspective. *Planning Theory*, 2(2), 101–123.

Hillier, J. 2010. Introduction: Planning at yet another crossroads?, in J. Hillier and P. Healey (eds), *The Ashgate Research Companion to Planning Theory. Conceptual Challenges for Spatial Planning*. Farnham: Ashgate, 1–34.

Mattila, H. 2002. Aesthetic justice and urban planning: Who ought to have the right to design cities? *GeoJournal*, (58), 131–138.

Needham, B. 2006. *Planning Law and Economics: The Rules We Make for Using Land*. Abingdon: Routledge.

Needham, B. 2007 *Dutch Land-use Planning*. The Hague: Sdu Uitgevers.

Ostrom, E. 2009. Design principles of robust property rights institutions: What have we learned, in G.K. Ingram and Y.-h. Hong (eds), *Property Rights and Land Policies*. Cambridge, MA: Lincoln Institute of Land Policy, 25–51.

Rittel, H.W. and Webber, M.A. 1973. Dilemmas in a general theory of planning. *Policy Sciences*, (4), 155–169.

Index

For Product Safety Concerns and Information please contact our EU
representative GPSR@taylorandfrancis.com
Taylor & Francis Verlag GmbH, Kaufingerstraße 24, 80331 München, Germany

www.ingramcontent.com/pod-product-compliance
Ingram Content Group UK Ltd.
Pitfield, Milton Keynes, MK11 3LW, UK
UKHW022133190926
475702UK00013B/1779